The Evolution of Agency

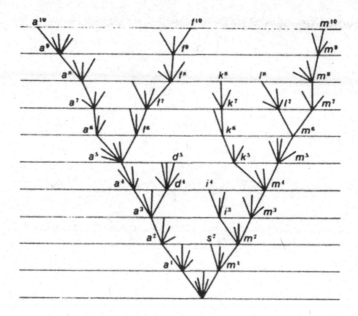

The Evolution of Agency

Behavioral Organization from Lizards to Humans

Michael Tomasello

The MIT Press
Cambridge, Massachusetts
London, England

The MIT Press would like to thank the anonymous peer reviewers who provided comments on drafts of this book. The generous work of academic experts is essential for establishing the authority and quality of our publications. We acknowledge with gratitude the contributions of these otherwise uncredited readers.

This book was set in ITC Stone Serif Std and ITC Stone Sans Std by New Best-set Typesetters Ltd. Printed and bound in the United States of America.

Illustrations by Marina Skepner.

Library of Congress Cataloging-in-Publication Data

Names: Tomasello, Michael, author.
Title: The evolution of agency : behavioral organization from lizards to humans / Michael Tomasello.
Description: Cambridge, Massachusetts : The MIT Press, 2022. | Includes bibliographical references and index.
Identifiers: LCCN 2021033956 | ISBN 9780262047005 (hardcover)
Subjects: LCSH: Evolutionary psychology. | Human behavior. | Animal behavior.
Classification: LCC BF698.95 .T65 2022 | DDC 155—dc23
LC record available at https://lccn.loc.gov/2021033956

10 9 8 7 6 5 4 3 2

Contents

Acknowledgments

I would like to thank several colleagues for helpful feedback on an earlier version of this manuscript. Josep Call read pretty much the entire first version and gave me extremely helpful feedback, as did Jan Engelmann, Walter Sinnott-Armstrong, and Manuel Bohn. I thank them all for helping to make this a better book. I also thank Brian Hare and Alex Rosenberg for a number of useful suggestions on parts of the earlier version as well. Further thanks go to Walter Sinnott-Armstrong and his seminar of colleagues (mostly from Duke University's Department of Psychology and Neuroscience) for a very interesting session devoted to the book that was helpful in numerous ways. I also thank Philip Laughlin at MIT Press and four anonymous reviewers for very helpful feedback. And finally, I express my deepest gratitude to my wife Rita Svetlova for help with both ideas and the manuscript itself, especially the introduction.

1 Introduction

In the distant future . . . psychology will be based on a new foundation, that of the necessary acquirement of each mental power and capacity by gradation.
—Charles Darwin, *On the Origin of Species*

Primates and other mammals seem to be more "intelligent" than smaller-scale creatures such as insects. But the basis for this impression is not at all clear. It is certainly not based on differences in the complexity of behavior: ants building anthills, spiders weaving spiderwebs, and bees communicating the location of nectar to hive mates are as complex as, or more complex than, anything any primate or mammal can do.

The issue is not complexity but control. Even when they are doing something highly complex, the behavior of ants and spiders and bees does not seem to be under the individual's control. Their evolved biology is in control. In contrast, even when they are doing something relatively simple, primates and mammals seem to be making active and informed decisions that are at least somewhat under the individual's control. In addition to their evolved biology, they are operating with a psychology of individual agency.

Individual agency does not mean total freedom from biology; it is always exercised in the context of an organism's evolved capacities. As just one example, it is clear that a squirrel is somehow preprogrammed to cache nuts. But the exigencies of a particular landscape at a particular moment are unique in ways for which the organism cannot be biologically prepared in detail, and so the individual squirrel, as agent, must appraise the current situation and make a caching decision for itself. For many organisms, the degrees of freedom in making such decisions are quite limited—and may differ in different behavioral domains—but such degrees of freedom nevertheless often exist, and within them it is the individual agent that decides what to do.

Evolutionary approaches to both animal and human behavior have, for whatever historical reasons, tended to ignore individual agency. Perhaps it raises the spectre of a homunculus behind the scenes that explains nothing. But biologists themselves faced a similar issue a century ago with the notion of an élan vital that purported to explain life in general. It turns out that living things are distinguished from nonliving things not by an animating substance or entity but rather by a special type of chemical organization. Similarly in the current case, we may say that agentive beings are distinguished from non-agentive beings not by an agentive substance or entity but rather by a special type of behavioral organization. That behavioral organization is feedback control organization in which the individual directs its behavior toward goals—many or most of which are biologically evolved—controlling or even self-regulating the process through informed decision-making and behavioral self-monitoring. Species biology is supplemented by individual psychology.

How and why did agency evolve, and why more so for some species (in some behavioral domains) than for others? A plausible hypothesis is that in some cases the environmental niche of a species is too unpredictable across space and time for hardwired perception-behavior pairings to be effective. In the face of such unpredictability, Nature—if we may personify the process of evolution by means of natural selection for ease of exposition (Okasha, 2018)—needs someone "on the ground," so to speak, to assess local conditions in the moment and decide on the best course of action. What thus evolves is an underlying psychology of agency that empowers the individual—in some key subset of situations—to decide for itself what to do according to its own best judgment. This way of operating represents an ancient organizational architecture characteristic of a large majority of extant animal species, and indeed, I would argue that even ants and spiders and bees make some individual decisions, even if they are few and highly constrained.

Agency is thus not about all of the many and varied things that organisms do—from building anthills to caching nuts—but rather about *how* they do them. Individuals acting as agents direct and control their own actions, whatever those actions may be specifically. The scientific challenge is to identify the underlying psychological organization that makes such individual direction and control possible. Answering this challenge yields a kind of photographic negative of the usual picture in evolutionary psychology: backgrounding what is usually focused (the adaptive specializations of species) and focusing what is usually backgrounded (the agency of individuals). To explain in the end specifically human agency—as I wish to do—we need an account that traces the evolutionary steps in agentive behavioral organization from creatures who make few and highly constrained decisions to creatures who quite often decide for themselves what to do. Perhaps surprisingly, it turns out that there are only a few such steps.

Evolutionary Approaches to Animal Psychology

From the beginning, Charles Darwin was interested in behavior. His Galapagos finches had beaks of varying shapes and sizes because—and only because—they had to *do* different things to get food. At one time or another, Darwin studied the behavior of dogs, cats, earthworms, an orangutan named Jenny at the London Zoo, barnacles, his firstborn child, and even climbing plants! In each case, he speculated about the underlying psychology involved, arguing for continuity via "descent with modification" (see the epigraph at the beginning of the chapter). Darwin also argued that the agency of individuals contributes to the variation needed for the evolution of behavior, and so played an important part in the process (Bradley, 2020). However, at that time, no organized scientific paradigm for the study of animal behavior existed that could turn Darwin's concern with behavior into a program of empirical research.

The first program of empirical research consistent with Darwin's original vision emerged only in the middle of the twentieth century. Konrad Lorenz, Niko Tinbergen, and Karl von Frisch founded the discipline of ethology, which focused on the evolved ("innate") behaviors of particular animal species. Their basic claim was that a species' behavior, just like its physiology, evolved as an adaptation to a particular ecological niche. Thus greylag goose mothers evolved innate behaviors to fetch their wayward eggs (or golf balls in experiments) and roll them back to the nest, and stickleback fish evolved aggressive behaviors in response to particular colorations on particular body parts of conspecifics. E. O. Wilson, in his book *Sociobiology* (1975), extended the approach to social behavior, including especially the highly complex social behaviors of eusocial insects such as ants and bees. The paradigm did not include much psychology—by design, as it dubbed itself "the biology of behavior"—and virtually no concern with individual agency.

In the past few decades, ethology has flourished, but under different names. It has basically transformed into what are now called behavioral biology and behavioral ecology. Like classical ethology, neither of these newer disciplines concerns itself with psychology per se (they are typically housed in departments of biology). Both of them focus on behavior, but only on the way that behavior contributes to genetic fitness. They sometimes refer to processes of decision-making, for example, in models of optimal foraging, but these are conceptualized not as psychological mechanisms controlled by individuals but rather as natural selection's way of shaping a species' behavior so as to maximize fitness benefits (e.g., caloric intake) and minimize fitness costs (e.g., energy expenditure). Potentially psychological or agentive terms like *choice* and *strategy* used in these analyses are thus only proxies for the evolutionary and genetic processes that contribute to behavior.

Although behavioral biologists and ecologists are mostly not concerned with the psychological mechanisms that generate behavior, psychologists are. The first psychologists with a systematic program of empirical research in animal behavior were the behaviorists, who actually began in the first half of the twentieth century, before the ethologists (by modifying earlier philosophical approaches to animal psychology). Behaviorists focused on one and only one thing—learning—in one or two species (first rats and then pigeons). They were not evolutionary psychologists. They took into account neither the ecological challenges nor the evolved behavioral capabilities of particular species—nor how species' evolved capabilities structure their learning—and were generally skeptical of the ethologists' claims of innateness (e.g., Skinner, 1966). Nor were the behaviorists cognitive psychologists: they explicitly eschewed any reference to "internal states" in their behavioral analyses (although they at some point came to allow memory for learned associations). Actively resisting both the evolutionary and cognitive revolutions in the study of behavior led to the demise of behaviorism in the late twentieth century. Nevertheless, some unhelpful remnants of the paradigm still survive in many areas of psychological research, especially the view of organisms as passive recipients of stimuli to which they reactively respond.

Another unhelpful remnant of behaviorism is the nature-nurture debate. If we are concerned with the psychological mechanisms by which organisms generate their actions, this is the wrong debate. The issue is not whether something is innate or learned, but rather the degree to which it is controlled by the individual. Thus an organism may have a genetically wired preference for sugary foods, but from the point of view of agency, the issue is whether this preference compels the organism to consume every sugary food it encounters, or whether this preference is merely one factor among several in the organism's individual decision of what to eat. In terms of cognition, some animals can use only one kind of tool in one delimited context, whereas chimpanzees can use a wide range of tools flexibly in a wide variety of contexts—including novel tools in experiments—and can even *make* tools to fit the situation when needed. Such behavioral flexibility based on individual judgment and decision-making does not of necessity involve learning; chimpanzees sometimes use novel tools flexibly upon first encounter. Rather, such behavioral flexibility emanates from a particular kind of behavioral organization that I am calling agentive.

The nature-nurture debate is rendered further moot when we recognize the artificiality of behaviorists' focus on a molecular level of punctate stimuli and responses. The behaviors of most organisms are enacted psychologically on multiple hierarchical levels simultaneously—a foraging trip is simultaneously seeking to satisfy hunger, searching for prey, traveling to a specific location, and moving limbs in certain ways—and some of these levels are more under the individual agent's control than others. A key way that the behavior of a species evolves is by the evolutionary emergence of new goal states that are more or less hardwired by Nature (e.g., an evolved preference for a new food), but with the behavioral means of achieving those goal states left up to the individual to figure out on its own (given its existing cognitive and behavioral capacities). This way of thinking about things recognizes—even in one and the same activity—the important role of both species-level genetic structuring and individual psychological agency.

By the late 1970s and early 1980s, many students of animal behavior had joined the cognitive revolution. By the 1990s there was a journal by the name of *Animal Cognition*, which published studies of a variety of animal species that mostly used the theories and methods of human cognitive science, including cognitive-developmental psychology. The emerging discipline included topics such as spatial cognition, object concepts and categories, the understanding of causality, the understanding of quantities, social cognition (theory of mind), social learning (imitation), communication, cooperation, and so-called horizontal skills such as memory and problem-solving. In studying such phenomena, animal cognition researchers mostly focused on those that are to some degree under the individual's control. Thus, in their overview of research and theory on primate cognition, Tomasello and Call (1997) explicitly stated that things such as spiders building spiderwebs are interesting and complex phenomena, but they are not psychological, precisely because they are mostly not under the individual spider's flexible control. The concept of agency thus, in a sense, represents the dividing line between biological and psychological approaches to behavior; it is the distinction between complex behaviors designed and controlled by Nature, as it were, versus those designed and controlled, at least to some degree, by the individual psychological agent.

Research in animal cognition has mostly focused on the various cognitive skills with which particular species operate. Much less research has investigated individual decision-making and behavioral control. In terms of decision-making, studies have established that some nonhuman primates employ some of the same decision-making processes as humans, including many of the "nonrational" biases discovered by human decision scientists, such as temporal discounting and loss aversion (Santos & Rosati, 2015; Mendelson et al., 2016). And in some cases, the ecological pressures leading to species' differences in styles of decision-making have been identified as well (e.g., Rosati, 2017a). In terms of behavioral control, studies have again shown that some nonhuman primates operate with some of the same processes of executive function as humans, and many of their ecological correlates have also been identified (Rosati, 2017b, 2017c). What is still missing in this work, however, is a systematic theoretical account of the evolution of individual decision-making and behavioral control. That is to say, what is missing is an account of how certain *types* of decision-making and behavioral control as instantiated in certain *types* of psychological architectures have evolved under certain *types* of ecological conditions, to enable individuals to make individual decisions. Ideally, this account would follow individual decision-making as it evolved from some ancient creatures to the present.

Evolutionary Approaches to Human Psychology

From Darwin's *The Descent of Man* (1871) onward, evolutionary explanations of the behavior and psychology of humans have met with active resistance from both scientists and nonscientists alike. There was especially vociferous protest against E. O. Wilson's attempt in the final chapter of his 1975 book to argue for an evolutionary basis for human social behavior. The resistance was based on a variety of concerns, but chief among them was that an evolutionary explanation equates to biological (genetic) determinism, which means that individual human agents are not responsible for their actions. This concern has been heightened by the rhetoric of Richard Dawkins (1976) and others to the effect that "selfish genes" are the actual causal agents, with organisms acting merely as their "vehicles."[1]

But science marches on. We now have an active scientific paradigm known as human behavioral ecology, practiced mainly by anthropologists (e.g., Winterhalder & Smith, 2000), which studies how humans living in (mostly) small-scale, traditional societies make a living and reproduce. As in behavioral ecological approaches in general, the focus is on the evolutionary, including genetic, bases of these activities, without particular concern for psychology per se. Focusing explicitly on psychology, John Tooby and Leda Cosmides (1992, 2005) have inaugurated a research program in human evolutionary psychology. They argue that, contrary to the assumption of many mainstream psychologists, the brain is not a general-purpose learning or computing device. Evolution does not just create generally useful mechanisms; it creates specific functional solutions to specific ecological challenges. Human psychology thus comprises a multitude of specialized, domain-specific computational mechanisms, each evolved to solve a specific adaptive problem, like the different blades of a Swiss Army knife. Searching for and choosing a mate are thus based on completely different psychological processes than searching for and choosing food. Most research in evolutionary psychology has focused on mechanisms with direct implications for survival and reproduction such as mate choice, kin identification, and cheater detection, as evolved in the genus *Homo* during the Paleolithic era, when humans were exclusively hunter-gatherers. These accounts have not been extended back systematically to any nonhuman animals. Although evolutionary psychology recognizes a variety of causes of human behavior and cognition, it has mainly focused on biological causes, and so it has also been cited for an excessive genetic determinism, to the neglect of the cultural dimensions of human psychology.

In an attempt to account for the cultural dimensions of human psychology, Peter Richerson and Robert Boyd (2005; see also Henrich, 2016) have developed a coevolutionary model of human behavior and psychology in which the individual inherits both its genes and its cultural environment, with a feedback loop such that individuals who are genetically adapted to function in a cultural environment do best (e.g., by having strong "social instincts" and skills of social learning). Because it incorporates culture in the process, this coevolutionary approach provides a richer starting point for investigating the evolution of specifically human behavior and psychology than does evolutionary psychology. However, like evolutionary psychology, it has not reached back systematically to nonhuman animals to determine how human behavior and psychology have evolved "by gradation" from those of other species. Research by my colleagues and myself (e.g., Tomasello et al., 2012; Tomasello, 2014, 2016) may be seen as an attempt to reach back at least to our great ape ancestors to discover how humans have evolved to create and acquire the cultural capacities that are so central to their functioning.

None of these evolutionary approaches to human behavior and psychology denies agency to human individuals. While these approaches stress that humans interact with the world via evolved cognitive and motivational mechanisms—and these influence to a greater or lesser degree the choices that humans make—nowhere in these accounts, as some critics have claimed, do we find a genetic determinism in which individuals are absolved of responsibility for their actions. Nevertheless, none of these approaches has focused specifically on the psychology of individual human agency. The psychological research that could potentially complement these evolutionary approaches is the study of human decision-making. However, the vast majority of this research is concerned with whether human decisions are normatively rational or subject to various nonrational biases (e.g., Kahneman, 2011). More to the point is research by Gerd Gigerenzer and colleagues (e.g., 1999, 2001) concerning the ways that humans actually make decisions, and how so-called nonrational biases are very likely evolutionary adaptations that help individuals to cope with risk and uncertainty in manageable ways (or helped them to cope with these in their evolutionary past). Research on human executive function and cognitive control is relevant and important as well (e.g., Egner, 2017), but to date there is almost no research comparing humans and other animal species.

The point is that there have so far been no systematic attempts to trace the roots of human decision-making and behavioral control deep into the evolutionary past. A systematic account of the evolutionary roots of human agency would require a starting point in humans' ancient animal ancestors well before primates, as well as a theoretical account that integrates processes of decision-making and behavioral control into more primal processes of goal-directed action. Attending to an extended evolutionary history before the emergence of modern humans creates a view of human psychology as a kind of layered onion, with an inner core of basic processes shared by all agentive organisms, further layers that humans share only with other mammals or primates, and an outermost layer of uniquely human psychology in all its dizzying complexity. Methodologically, seeing through to the functioning of the ancient inner layers of human psychology is difficult or impossible by studying only mature adults, for whom these inner layers are buried deep inside culture, language, and self-consciousness. It might be advisable to begin, therefore, by looking at the relatively simpler psychological functioning of relatively simpler organisms, as representative of humans' ancient ancestors. Says Aristotle in his *Politics*: "He who considers things in their first growth and origin . . . will obtain the clearest view of them."

Goals of the Book

My goal in this book is to reconstruct the evolutionary pathway to human psychological agency. Whereas the number and variety of specific behavioral adaptations across animal species are immense, the psychological mechanisms by which individual agents direct and control their behavioral decision-making are limited in number. Some very ancient human ancestors, such as the earliest bacteria, are not psychological agents at all—their behaviors are neither directed at goals nor individually controlled—and some agentive creatures such as birds and bees are off the evolutionary line to humans and thus not considered here. On the evolutionary line to humans specifically, I propose four main types of psychological agency— four schemes of organizational architecture for individual decision-making and behavioral control—in four taxa representative of important human ancestors. They are, in evolutionary order of emergence: goal-directed agency in ancient vertebrates, intentional agency in ancient mammals, rational agency in ancient great apes, and socially normative agency in ancient human beings.

To accomplish this evolutionary reconstruction, the most pressing need is a theoretically coherent and widely applicable model of the organizational architecture of agency, including specification of the key elements that must be added or transformed to go from simpler to more complex forms. My secondary goal, therefore, is to provide a simple but comprehensive model of agency that, with appropriate modifications, is applicable across a wide spectrum of animal behavior from humans' most ancient animal ancestors to modern humans. Such a model must perforce include the perceptual and cognitive capacities that are necessary for an individual of a given species to make the behavioral decisions it needs to make and, in addition, to self-regulate the process as it unfolds over time. Because agency is not just another specialized behavioral or cognitive skill, but rather the most general organizational framework within which individuals formulate and produce their actions, tracing the evolutionary roots of human agency amounts to nothing more or less than an evolutionary explanation of human psychological organization in general. Such an account will require both a broadening and a deepening of current theories of evolutionary psychology.

2 A Feedback Control Model of Agency

Natural selection is the theory of how forms come to be adaptive, that is, to be governed by a quasi purpose. It suggests a machinery of efficiency to bring about the end.
—Charles Sanders Peirce, *Collected Works*

All animal species engage in a variety of biological activities, including self-produced movements, that promote their survival and reproduction. For some scholars, these activities make all organisms "adaptive agents" in the evolutionary process (e.g., Walsh, 2015). This organismic perspective is extremely valuable in current discussions on the nature of evolution, which often focus only on molecular-level processes. Nevertheless, this approach to agency is both too broad and too biological for current purposes. My focus here is rather on the more circumscribed notion of what might best be called psychological agency (see also Sterelny, 2001).

Behaving as a psychological agent means that the underlying psychological processes that generate actions are organized in a particular way. An agent does not just respond to stimuli but actively *directs* (or even plans) its actions toward goals, actively attending to relevant situations in order to do so. And an agent does not just "aim and shoot" at its goals ballistically but rather flexibly *controls* (or even executively self-regulates) its actions by making informed decisions about what will work best at various points in a dynamically unfolding situation. Methodologically, the main evidence for psychological agency is the "behavioral flexibility" of individuals, especially in novel circumstances (see Lea et al., 2020, for discussion of the importance of this concept for modern research in animal cognition).

Behavioral flexibility suggests that the individual organism is finding new ways in the moment to deal with challenging new circumstances.

If the goal is to reconstruct the different forms of psychological agency on the evolutionary pathway to humans, I must first do three things. First, to characterize the several forms of agency rigorously, I need to find or devise an integrated set of theoretical tools for describing the organizational architecture of agency. The theoretical tools of behavioral ecology are not geared to this task; those of behaviorism, such as they are, are too narrowly focused on processes of learning and memory; and those of evolutionary psychology are too focused on modularized skills to the neglect of over-all psychological organization. The theoretical tools of animal cognition, supplemented by those of coevolutionary theory for humans, are appropriate to the task but are severely underdeveloped. My attempt, therefore, is to develop the theoretical tools of animal cognition (including humans) by adapting and extending a class of psychological models of agency from modern cognitive science. The basic structure of psychological agency, I contend, is manifest in classic cybernetic models of goal-directed action based on principles of feedback control (e.g., Miller, Galanter, & Pribram, 1960). Then this basic structure must be fleshed out using models from modern decision science focused on different types of decision-making under different types of uncertainty (e.g., Gigerenzer et al., 2011). Then for some organisms we also need models of executive function and cognitive control to provide additional resources for characterizing the way that individuals act intentionally and self-regulate their actions from an executive tier of functioning (e.g., Egner, 2017). Then, for still other cases, we need to adapt models of metacognition and "computational rationality" in which agents assess the efficiency of their first-order executive processes from a second-order executive tier (e.g., Gershman et al., 2015). Finally, for humans we need to transform everything to characterize their unique forms of socially shared agencies, which change fundamentally the way that individuals make and self-regulate their decisions.

The second thing I must do as preparation for my reconstructive task is to identify the precipitating ecological circumstances that serve to explain evolutionary transitions from one form of agency to another. On the basis of general theoretical considerations from decision science, of particular importance are the situations of uncertainty—indeed, the different types of uncertainty—that decision-makers face (e.g., Yu & Dayan, 2005). The evolutionary hypothesis is that when individuals regularly face situations of uncertainty, the individuals that fare best are those that operate agentively to flexibly assess the situation at hand and make a decision informed by the relevant local contingencies (perhaps with the aid of some heuristics; see Gigerenzer et al., 2011) and then monitor and self-regulate behavioral execution as it unfolds. My more specific evolutionary hypothesis is that the four main *types* of agentive organization on the way to contemporary humans evolved in response to four main *types* of uncertainties, created mainly by four different *types* of social interaction. I identify these types of uncertainty using a procedure that is the opposite of reverse engineering, what might be called prospective engineering. Instead of starting with the mechanism and attempting to determine what problem it evolved to solve, I start with the problem (ecological challenge) and attempt to determine—on the basis of empirical observations from behavioral experiments—what mechanisms might have been designed to solve it.

Third and finally, because we cannot observe the behavior of extinct creatures, if we want experimental information about the behavior of humans' ancient ancestors, we must identify extant creatures that can serve as model species. For this task, I use the normal methods of comparative biology, namely, gleaning general information from fossils about the physiology of the key extinct organisms and their evolutionary trajectories, and then using this information to find extant organisms to use as models. We can then have access to data from behavioral experiments with these model species.

Machine Models of Agency

Psychologists have always been attracted to machines as models for how behavior is generated and organized. Classical behaviorists used as a model the automated telephone switchboard from the 1930s. The switchboard was quiescent until a call came in (the stimulus), at which point it connected the phone line of the caller to the phone line of the dialed number (the response—which, if reinforced by success, was learned). The ethologists had a different machine model, but it was just as passive. In Lorenz's hydraulic reservoir model, the organism had certain "action-specific energies" that built up over time (e.g., hunger). Then along came an innate releasing mechanism (e.g., the sight of food as stimulus), which activated some fixed action patterns (e.g., consummatory behavior as response), which drained the hunger reservoir. This passive view of animal behavior continues today, at least implicitly, as many scientists continue to speak of behavior as a response caused by a stimulus (see addendum A). And reinforcement learning, a theory popular in the computational modeling of human behavior, has organisms passively reacting to "reward signals" in the environment (see Juechems & Summerfield, 2019).

Switchboards and reservoirs both operate in the manner of physical causality: a linear process in which cause (stimulus or releasing mechanism) leads to effect (response or fixed action pattern). But the living world operates more actively, even proactively. The basic bodily processes that maintain life are organized into a circular causality, homeostatically, in which there are internally represented reference values that the body actively seeks to attain or maintain. In mammals, for instance, the body works actively to maintain a constant temperature in its internal environment despite external perturbations. Following Wiener (1948), Ashby (1952), and other early cyberneticians, Miller et al. (1960) proposed that organisms' behavioral interactions with the environment are organized circularly in the same manner: the organism has goals, which it actively pursues via behavioral plans aimed at goal attainment, along with perceptions that provide feedback about how the behavioral plan is working. The basic unit of behavior is thus not a linear and passive stimulus-response pairing, but rather an active and circularly organized feedback control mechanism. In modern cognitive science, this is the standard model at the heart of all computational models of intelligent action and cognition (e.g., Gershman et al., 2015). Examining how machines organized as feedback control systems operate—and how this might relate to the organization of behavior and psychology—thus offers a useful starting point for an evolutionary account of human agency.

We can clearly see the basics of feedback control organization in an HVAC (heating, ventilating, and air-conditioning) system controlled by a thermostat (fig. 2.1). The typical HVAC system actually comprises two systems, a furnace for heating and an air conditioner for cooling, and the human flips a switch to decide which one is operating. The goal is to maintain the temperature of an indoor space at a constant level: heating the air when the outside temperature tends to make the inside space cooler (in winter), and cooling the air when the outside temperature tends to make the inside

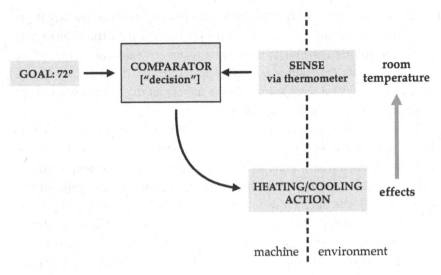

Figure 2.1
Basic feedback control organization of an HVAC system. Light-gray upward-pointing arrow is the feedback from action effects to perception.

space warmer (in summer). In the case of the furnace, the process begins when a human flips the switch to heating and sets a desired temperature. Using a thermometer of some sort, the thermostat then senses the actual room temperature and compares it to the desired temperature. If the room is colder than the reference temperature, the thermostat turns on the furnace (itself a complex, multicomponent machine). Flipping to the air conditioner activates a similar process in the other direction: when the air is warmer than the reference temperature, the thermostat turns on the multicomponent air conditioner. Some HVAC systems integrate the two functions by sensing whether the room temperature is either above or below the set value, and turn on the furnace or air conditioner as appropriate. In this case, rather than the go-no-go decision of the typical thermostat ("deciding" whether to turn on the furnace or not), the more complex HVAC system makes either-or decisions about which action to choose according to what is needed to meet the goal.

All autonomous, "intelligent" machines have this same circular causal organization: action (e.g., turning on heat) causes change in perception (e.g., sensed temperature), which is then compared to the reference value or goal (e.g., 72 degrees) to determine if further action is needed. This circularity contrasts starkly with the linearity of electric fans or space heaters, which are simply turned on or off by a human (i.e., the human acts as the controlling thermostat by sensing room temperature and deciding what is needed). Decision-making typically operates in an HVAC system via a simple physical mechanism, often a metal coil that simply expands or contracts as a function of temperature.

To see the consequences of this kind of behavioral organization, let us engage in an act of prospective engineering. Imagine that I have the goal of keeping my lawn free of leaves. One thing I could do to get help with the task is to use a vacuum to suck them up; the vacuum would perform the action while I, the operator, would supply the goal, direction, and perceptual feedback. The machine is just a tool. But I could also build a machine to act by itself. For example, I could make a mobile leaf vacuum machine that engaged in a random walk all over the lawn (perhaps constrained by fences all around), which sucked up everything it encountered into a hopper. This would work but would be highly inefficient, as the machine would frequently wander over empty lawn with its motor on, wasting battery power. So perhaps I could add a camera enabling the machine to "see" leaves and react: the camera sees a leaf at a particular location (stimulus), which activates the locomotory apparatus to go to that location and suck up the leaf (response). But such an open-loop, stimulus-response machine would have no way of making adjustments en route: if the wind blew away the leaf at which the machine was aiming just before it got there, it would go there anyway. Or if a branch fell in the way, the machine would just bang up against it endlessly. And when the hopper filled up with leaves, the machine would just keep sucking fruitlessly, leaving leaves on the lawn, with no way of determining when to stop. What we need is a leaf vacuum machine that operates as a feedback control system or, even better, as a hierarchy of interrelated feedback control systems (or a heterarchy as a variant on this organizational scheme; see Bechtel & Bich, 2021).

Figure 2.2 depicts a highly simplified diagram of how a hierarchically organized feedback control leaf vacuum machine might work, highly simplified because each of the four components represents its own hierarchically organized feedback control system (i.e., each contains further levels of implementation, such as tuning on motors, aligning wheels, etc.). Each component has a separate mechanism with a goal (G), and it acts (A) until its perception (P) matches its goal. If there is no match, it keeps trying; and

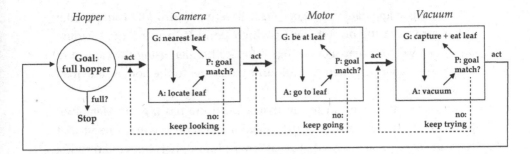

Figure 2.2
Highly simplified sequence of feedback control systems in an autonomously operating leaf vacuum machine that cleans a lawn efficiently and flexibly. *G* = goal; *A* = action; *P* = perception (to see if actual situation matches goal situation). Each box represents a hierarchy of submechanisms (e.g., turning on motor to move, aligning wheels, etc.).

if there is a match, it passes things off to the next component to do its job. The components are organized and connected such that the lower-level goals operate in the service of the higher-level goals: the machine only locates a leaf so it can travel to it; it only travels to it so it can eat it; and it only eats it so it can fill the hopper (at which point a human, reading some signal, must intervene to empty the hopper and set the machine off anew). If, as it is moving toward a leaf, that leaf blows away unpredictably, then the machine adjusts to the new situation, for example, by locating the now nearest leaf as new goal. If an obstacle drops in the way en route, then, with a small programming tweak, the machine could just abort by sensing that the obstacle is a dead end that requires stopping and then aiming at a new leaf. Compared with the random-walk and stimulus-response leaf vacuum machines, this feedback control leaf vacuum machine is much more flexible and efficient in getting the job done.

I have skimmed over or omitted many details here, but the general point is clear: the best way to get the job done efficiently and flexibly is to use a system that has goals and pursues them with perceptual feedback along the way. For tasks with any complexity, the best system is one that operates with a hierarchy of feedback control components, for example, in vacuuming leaves: (i) a hopper that senses when it is full, and sends a call to action when it is not; (ii) a camera that senses the presence of leaves and orients the machine toward the nearest one and activates locomotion; (iii) a locomotory apparatus that goes until it senses (using the camera) that it has arrived at the nearest leaf, at which point it turns on the vacuum; and (iv) a vacuum that sucks things into the hopper, using the camera to check its success. And then control turns back over to the hopper, and it all starts again.

But what does it mean to say that a machine has a goal? Metaphysics aside, it simply means that the designer has built the machine so that there are perceptual images that act as what philosophers call pro-attitudes: perceived states of affairs that the actor is "motivated" to bring about by behaving. One might object to using this kind of language for machines—surely our leaf vacuum machine does not have goals—but it is useful precisely because the designers of such machines are trying to build them to do the job the way an agentive organism would do it. A human would have a desired image of the lawn, and then clean the lawn until she perceived that the actual state of the lawn matched that image, and then stop (Powers, 1973). So let us just operationally define goals in this way as desired perceptions—but with one clarification. Consider breathing. We do not normally think of breathing as a goal-directed activity because we seemingly just do it. But if someone is deprived of oxygen, they will immediately begin engaging in goal-seeking behavior. Humans have a reference value of a constant supply of oxygen, which breathing unthinkingly supplies, and when that reference value is disrupted (e.g., underwater), they react by pursuing the behavioral goal of returning to the steady state. Or consider a father in the yard with his daughter. His reference value is that she stay in the yard. So when she stays in the yard, he does nothing. But if the child strays into the street, goal-directed behavior (fetching her back into the yard) ensues. So, in this analysis, even doing nothing is goal-directed in the sense that it is what one needs to do to maintain a certain reference value. The machine or organism acts so as to bring or keep its perceptions in line with its reference values, which in some cases means behaving and in other cases means doing nothing.

If we think about our leaf vacuum machine as an organism, it is important to point out that its highest-level goal is to fill its hopper with leaves (satisfy its "hunger" for leaves). But the human employing the machine has even higher-level goals. For example, I may be removing the leaves because I want a beautiful lawn as a way of impressing my neighbors. That is why I built the machine in the way that I did: to do the job in a way that fulfills my higher-level goals. But the higher-level goals are not part of the leaf vacuum machine's psychology, if you will. In the next chapter, I analogize this situation to situations in which Nature, metaphorically speaking, wants the organism to behave in a certain way for her own goals, metaphorically speaking, but the organism knows nothing of these goals. For example, Nature wants the organism to survive, but the individual creature has no access to this larger evolutionary goal; its psychology only includes eating, escaping predators, and so on.

The feedback control model of behavior thus comprises a hierarchy of systems, each with three central components: (i) a reference value or goal, (ii) a sensing device or perception, and (iii) a device for comparing perception and goal so as to make and execute a behavioral decision.[1] It is not just that a feedback control system happens to be a good model for generating flexibly intelligent, agentive behavior; *it is the only type of model possible.* Consider one further thought experiment. You are frustrated by an inefficient traffic light, at which you wait when there are no other cars in sight, and which turns green for short lines of traffic when longer lines of traffic are waiting at other incoming roads. One day, all of a sudden, the traffic light is working as efficiently as a traffic policeman. What might the city planners have done to that traffic light? It is almost impossible to imagine what they could have done other than supply the traffic light with (i) some representation of the ideal goal state for the traffic to be in (e.g., equal lines all around); (ii) some form of perceptual access to the current traffic situation (e.g., cameras); and (iii) some decision rules for turning the lights on and off to change the perceived traffic situation in the direction of the goal state. What else could it be? For certain, that is how a traffic policeman does it.

Our basic model of agency is thus a feedback control system in which the individual directs its actions toward goals and controls those actions via perceptually informed decisions. Thermostats and leaf vacuum machines are specifically designed to be goal-directed, but it is not clear that it is appropriate to say that they make decisions, since the options available to them are built in by a human and effected more or less mechanically. But we do not need to answer the basic philosophical questions of artificial intelligence here. I will thus defer discussion of decision-making until we get to living organisms. I will also defer for now discussion of another potentially significant component of agency: executive function and cognitive control.

Many feedback control mechanisms employ an executive tier of function-ing (and even possibly a second-order executive tier on top of that) that supervises the operational level of perception and action. But, again, the complexities involved may be dealt with more effectively in the context of the behavior of living organisms.

Types of Ecological Challenges

To explain the evolutionary transition from one form of agency to another, the general formula, as in all evolutionary explanations, is to begin by identifying the ecological challenges that made the earlier form less viable and the newer form more viable. Drawing from human decision science (e.g., Yu & Dayan, 2005), I propose that agentive organization arises when organisms are regularly faced with one or another type of uncertainty, for example, risk (the probabilities of potential outcomes are known), ambi-guity (the probabilities of potential outcomes are unknown), and volatil-ity (the probabilities of potential outcomes change unpredictably during action execution).

Many current discussions in the field of animal behavior focus on whether the relevant ecological conditions emanate mainly from the physical—usually foraging—environment or from the social environment. Thus, for the past few decades researchers have debated whether the evolu-tion of "intelligence" in primates and other animals—often operational-ized as one or another measure of brain size—is due to ecological factors, operationalized as something like area of foraging range, or social factors, operationalized as something like social-group size. But this way of looking at things is grossly oversimplified (as most of the participants in this debate explicitly recognize).

The first oversimplification is the construct of "intelligence." This is a concept invented by human psychometricians a century ago to predict which human children would benefit from higher education; it has noth-ing to do with the evolution of cognitive processes in nature, which are myriad and more or less specialized for particular functions (Rosati, 2017a). Tomasello and Call (1997) note that foraging, to single out just one func-tion, involves at the least the following subfunctions and their respective adaptations: finding food requires skills of spatial cognition; identifying food requires skills of object categorization; quantifying food requires skills of quantification; and using tools, for those species that do, requires skills of object manipulation and, potentially, causal understanding. In the social domain, different adaptations are required when individuals compete with others (e.g., for food or mates), exploit the knowledge and skills of others (e.g., by following their gaze direction or socially learning from them), and cooperate with others (e.g., in coalitions and alliances). In all these various domains, different species may have some of these skills but not others, or have these skills in differing forms or degrees; there is no single general trait called "intelligence."

The second oversimplification is that ecological complexity and social complexity are not adequately operationalized by size of foraging range and size of social group, respectively. Moreover, ecological and social factors often interact with one another in complex combinations, as many species forage in social groups. My approach here is to focus on the major source of uncertainty in the lives of organisms—namely, other organisms—but in more complex ways than any simplified measure of "social complexity." Specifically: (i) for creatures who either hunt or are hunted, the unpredictabilities of prey and predator behavior are key; (ii) for creatures who compete with group mates for food in scramble competition (to the swiftest go the spoils), the key is efficient decision-making about ecological conditions so as to win the race; (iii) organisms who engage in contest competition with group mates over resources may develop even more efficient decision-making skills as well as special skills of social cognition for better predicting the behavior of their competitors; and (iv) individuals who obtain their resources mainly by collaborating with conspecifics may evolve skills for predicting and controlling the behavior of collaborative partners who have the same basic psychology as themselves. In general, my proposal is that while foraging ecology is key for the evolution of many cognitive skills in areas such as space, object categories, and quantities, the evolution of agentive decision-making occurs mainly in response to different types of unpredictabilities, and we can have little doubt that the most unpredictable entities in the lives of most organisms are other organisms. The outcome is a hypothesis in which social and ecological factors and the resulting behavioral adaptations are intermixed in complex ways (e.g., the organism makes better and faster foraging decisions because of social competition).

A final important point: Acting as an effective decision-maker requires perceiving and understanding both the physical and the social environment in relevant ways, and this is driven by the actions that need to be performed. For example, our leaf vacuum machine is built to perceive leaves, and only leaves (e.g., ignoring discarded bottles from a party last night), because this is what the machine needs to achieve its goals as specified. As a biological example, some Galapagos finches have the visual capacities to detect camouflaged insects, whereas others detect the ripeness of certain species of nuts, depending on their respective foraging activities. These are all specific forms of perception for specific functions. But in the case of agency, as a type of psychological organization, the key point is that changes in the agentive organization of action lead to changes in the *types* of things the agent may experience. Thus, for example, an organism that executively self-monitors its own behavioral and psychological functioning is privy to a dimension of experience not available to organisms that do not engage in such self-monitoring. Therefore a further dimension of agency that will play an important role in my account is what I will call the *experiential niche* of the organism, as driven both by its particular adaptations to its particular ecological niche and also, on a more general level, by the nature of its overall agentive organization.

Extant Species as Models for Extinct Species

Our question is how human agency evolved, in terms of the various evolutionary steps leading to it. In principle, we could answer this question by examining the history of agentive organization from humans' most ancient animal ancestors up to contemporary humans. But, of course, such a history is impossible in the sense that behavior does not fossilize (nor does brain tissue). But behavior does show continuity over time across related species, and this suggests the basic strategy of comparative biology: searching for the roots of human agency in contemporary species that are generally representative of some of humans' various animal ancestors. From among extant species, we may choose particular model species (or classes of species) on the basis of their hypothesized similarity to ancestor species at key evolutionary junctures.

Importantly, I will not focus here on cases of parallel or convergent evolution in agentive organization, for instance, in birds or eusocial insects. Rather, I will restrict myself to the evolutionary line leading to humans. Nor will I follow the analytic strategy of Peter Godfrey-Smith (2016, 2020), who examines many varieties of animal psychology or consciousness, including for some extremely simple creatures, with an eye to establishing the most basic modes of functioning common to all animal species. Rather, I will attempt to reconstruct an actual evolutionary sequence based on homologies, starting from simpler forms of behavioral organization in humans' ancient animal ancestors and then proposing how these successively transformed into more complex forms of behavioral organization, each building on its predecessor as an adaptive response to new ecological challenges (see Bonner, 1988, for a compelling theoretical account of how complexity evolves). The result is thus not a *scala naturae* of contemporary species but rather a reconstructed evolutionary sequence for which I could, in principle, choose any extant species as my end point. As a psychologist, I choose humans.

I have chosen particular species as representative of each of the four posited types of agency partly on principle and partly on expediency. As representative of the first animate actors (urbilaterians), who were not really organized agentively, I have chosen *C. elegans*, a wormlike creature about whom much is known. As representative of the first goal-directed agents, I have chosen lizards and other reptiles, with whom many behavioral experiments have been conducted. As representative of the first intentional agents, I have chosen squirrels (and their cousins, rats) and other mammals, again with whom many behavioral experiments have been performed. As representative of the first rational agents, I have chosen chimpanzees, the great ape with whom the most behavioral experiments have been performed. And as representative of the first socially normative agents, I have chosen early humans at two different evolutionary time points (based both on archaeological evidence of behavior and on an analogy to human children, with whom many behavioral experiments have been performed). My claim is that these species are as reasonable extant representatives of the types I am targeting as any, and indeed I was inspired in my choices by descriptions in the literature that the first organisms with nervous systems were wormlike, the first vertebrates were lizard-like, and the first mammals were squirrel-like. Chimpanzees are as good an ape as any to represent early great apes, and for early humans, we can make some analogies to contemporary human children. (See addendum B for some further information about research in animal cognition and how it could be supplemented and improved for these kinds of analyses.)

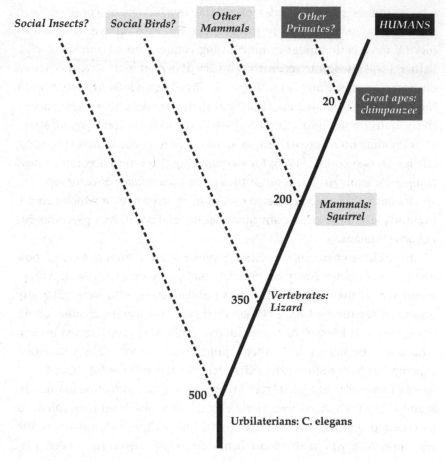

Figure 2.3
Evolutionary tree locating the species on which I focus here, namely, the taxa and
species in the rightmost margin: urbilaterians, lizards, squirrels, chimpanzees, and
humans. Nonagents are in roman type; goal-directed agents are in italics; intentional
agents are on a light-gray background; rational agents are on dark-gray backgrounds;
and normative agents are in all caps with black backgrounds. All the left-directed
(dashed) branches are species not covered, with some speculations about their agentive status (indicated by style of type), in some case via processes of parallel evolution.

In any case, figure 2.3 provides an extremely general depiction of the evolutionary tree within which we may locate my chosen model species. The chosen model species are in all cases on the rightmost (solid) branches leading to humans (with the approximate date of first branching indicated in millions of years ago). Nonagents appear in roman type; goal-directed agents are in italics; intentional agents are in italics on a light-gray background; rational agents are in italics on a dark-gray background; and socially normative agents are capitalized on a black background. The figure also depicts a few speculations about instances of parallel evolution off the human line. As mentioned briefly in the text, some insects may have evolved in parallel with reptiles to be goal-directed agents, and some birds have evolved in parallel with mammals to be intentional agents. Interestingly, and in line with my hypothesis about instigating ecological conditions, these potential cases of parallel evolution seem to occur most often in highly social species.

I am under no illusions about the arbitrariness of my choices for model species; no perfect solution presents itself. But given my hypothesis that the general organizational architecture of agency comprises only a few basic types and they are highly conserved across related species, I believe that the resulting picture is more or less accurate.

3 Ancient Vertebrates as Goal-Directed Agents

The individual organism determines in some sense its own environment by its sensitivity.

—George Herbert Mead, *Mind, Self, and Society*

Natural selection operates only on what it can "see," and that is the organism's overt actions (and its physical body as adapted for actions and other adaptive functions). The underlying psychological processes organizing and generating actions are thus naturally selected, but only indirectly through their effects on action. If an organism is capable of performing an action but does not perform it, that capability cannot become a target of natural selection. If an organism is motivated for an action but does not perform it, that motivation cannot become a target of natural selection.

Analogously, scientists infer psychological processes from their effects on the organism's overt actions. (The qualifier *overt* distinguishes the current approach from the "basal cognition" approach, in which all organismic functions, including bodily maintenance functions, are considered to be cognitive because they involve information processing; e.g., Lyon et al., 2021; Keijzer, 2021.) Scientists infer psychological agency when the organism acts flexibly toward its goal even in novel contexts. To behave in this flexible manner, the individual must go beyond a stimulus-driven, one-to-one mapping between perception and action. The individual must be capable of choosing to act or not to act, or among multiple possible actions, according to its continuous perceptual assessment of the situation as it unfolds over time (sometimes employing executive processes such as inhibition, as a further control process, during action execution). Animal species who actively behave in the world, but not in these psychologically agentive ways, may be called animate actors.

Animate (Nonagentive) Actors

The first organisms on planet Earth were not psychological agents. They did not need to be; they came into existence literally swimming in food. They were unicellular organisms that simply moved around with "open mouths." This interpretation is supported by the fact that similar unicellular creatures alive today do not act to pursue and consume nutrition and then stop when sated (indicating goal pursuit and satisfaction); they constantly move and consume more or less as "filter feeders." Their "decisions" are mechanical, and indeed, they do not even have separate sensory and decision-making mechanisms, only molecules sensitive to nutritious and noxious chemicals, leading automatically to certain actions. Such simple creatures cannot decide *not* to move toward nutritious chemicals, even when they are already sated, and they cannot make connections between their actions and the results to determine success. They are stimulus driven, not goal-directed (see Yin & Knowlton, 2006). The goals of eating nutritious things and avoiding noxious things belong to Nature, as it were, not to the individual.

Then, more than 500 million years ago, there emerged a wormlike creature that was the first human ancestor to operate with a nervous system: the urbilaterian. We know what it looked like because we have recovered an almost intact fossilized individual, and we can infer that it had a nervous system by looking at the genetic development of a wide range of similar

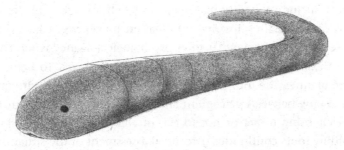

Figure 3.1
Imagined early urbilaterian about 500 million years ago.

species. From a behavioral standpoint, our current best guess is that the urbilaterian was capable of "mobility and sediment displacement" (Heinol & Martindale, 2008). Perhaps its behavior was similar to that of an extant creature about whom we know a great deal, *C. elegans*, a wormlike creature that serves as a model species in behavioral biology. All 302 of its neurons—many of which are clustered in ganglia, and 32 of which serve a chemosensory function—have been identified, along with all their synaptic connections. Not only do the chemosensory neurons detect either good or bad things and "signal" the motor neurons to produce bodily contractions that propel the organism either toward or away from those things, but *C. elegans* also uses the rate at which it is ingesting food, typically bacteria, to detect the location of richer and less rich clumps (Scholz et al., 2017). Moreover, if a behavior such as forward movement brings a bad result (e.g., a noxious chemical), the creature can perform one of two actions to move away (Hart, 2006). *C. elegans* finds its food by moving around in its environment actively, sometimes even learning the location of food in novel environments after several encounters (Qin & Wheeler, 2007).

The behavior of *C. elegans* would thus seem to be organized in a more complex manner than that of unicellular organisms. They have different mechanisms for sensing things in the world and acting in response. Classically, the function of a nervous system is to connect separate mechanisms of perception and action, and ganglia are seats of this integration, so it would seem that the separate mechanisms of perception and action are integrated in *C. elegans* (and also, by inference, in early bilaterians). However, it is unlikely that there is also a comparison with some kind of internal goal to create direction: their locomotion is mostly random or stimulus driven (Scholz et al., 2017). And these organisms do not seem to exhibit anything that we would want to call behavioral control: they do not inhibit or otherwise control action execution, and what they learn is simply the location toward which to direct their hardwired movements. It is thus unlikely that early bilaterians, as modeled by *C. elegans*, were goal-directed, decision-making agents, only animate actors.

Goal-Directed Agents

Sometime after the first bilaterians, perhaps about 500 million years ago, came the so-called Cambrian explosion and organisms with "complex active bodies" with appendages, teeth, claws, and more that had to be coordinated for effective action (Godfrey-Smith, 2016). These new complexities were adaptations to a much more complex and unpredictable set of ecological challenges. Organisms' foraging for food became much more uncertain as they began to prey on other highly mobile creatures who could flee or otherwise defend themselves, and they also had to defend against clever predators. (Continuing our exercise in prospective engineering, imagine that our leaf vacuum machine faced a lawn full of leaves, each of which could deploy a number of clever escape or attack strategies.) To cope with these many and varied challenges, organisms needed a much more complex manner of functioning than *C. elegans*. They needed not only a larger arsenal of appendages and actions but also more effective ways of controlling their actions flexibly to solve problems in uncertain and dynamically changing circumstances. Enter feedback control organization. These were the first truly agentive organisms, at least in some components of some domains of activity.

We do not know who the first agentive organisms were, but let us fast-forward to some creatures about whom we know much more. From the mélange of species with complex, active bodies emerged the first vertebrates, the fishes, and then some 350 million years ago the first land-based vertebrates, the amphibians, and then the reptiles. Because we have much

Figure 3.2
Imagined early vertebrate 350 million years ago.

better behavioral data on reptiles, let us focus our attention on them. On the basis of the fossil record, the first reptiles were twenty to thirty centimeters long, lizard-like creatures with legs and teeth and eyes. They also had relatively large brains. By all indications, they lived mainly by eating insects. If extant reptiles give any indication, these early reptiles' behavior was quite complex and flexible, especially in foraging. Of course, the ancient ancestors are long gone, but let us look in a general way at the behavior, especially foraging behavior, of some extant species of lizard.

In comparison with *C. elegans* and other worms, the foraging behavior of lizards is strikingly flexible across time and space. Foraging behaviors exhibit great variability across seasons, depending on the seasonal availability of different species of insects and spiders. And depending on the prey, at least some species of lizard have the option to choose between a sit-and-wait (ambush) strategy and a more active pursuit strategy. In the laboratory, systematic tests show at least some variability among individuals due to learning, especially in their foraging behavior. For example, lizards can learn quickly to solve novel foraging problems, such as removing the lid from a plastic well to retrieve a reward (Leal & Powell, 2012). In a variation on this theme, Qi et al. (2018; see also Szabo, Noble, Byrne, Tait, & Whiting, 2019; Szabo, Noble, & Whiting, 2019) required their lizards to discriminate between different lids on wells—only some of which had rewards—before choosing and removing one, which the lizards learned to do relatively successfully. Lizards also show flexible, context-sensitive behavior in their predator escape, as they escape in different ways from different experimentally structured threats (Cooper et al., 2007). Overall, Wilkinson and Huber (2012, p. 141; see also Szabo et al., in press) conclude in their general review of reptile cognition that "there is evidence of efficient learning in the spatial, physical, and social domains as well as examples of behavioral flexibility in food-acquisition tasks."

All this behavioral flexibility and learning, both within and between individuals, suggests an organization of behavior that is not just stimulus driven but directed at goals and controlled by individual decisions. Figure 3.3 depicts a simplified behavioral hierarchy of a lizard foraging for ants (suspiciously similar to our leaf vacuum machine). The hungry lizard emerges from its burrow, searching for prey (or the prey's likely habitat), goes to it and searches or waits until prey comes within reach, and then captures and consumes it. (Again, each of these four components contains

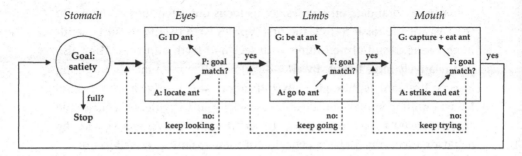

Figure 3.3
Highly simplified sequence of feedback control systems comprising a lizard's foraging for an ant efficiently and flexibly. *G* = goal; *A* = action; *P* = perception (to see if
actual situation matches goal situation). Each box actually represents a hierarchy of
submechanisms (e.g., moving limbs to locomote, opening mouth to eat, etc.).

further internal levels of implementation, such as moving limbs in certain
ways, etc.) In terms of learning, lizards go beyond *C. elegans* by engaging
in discrimination learning: learning to pursue a goal in one way in one
perceptual situation and in another way in a different perceptual situation. But, importantly, in discrimination learning, the individual does not
learn new behaviors but only learns which perceptual situations are most
rewarding using existing behaviors. Thus Suboski (1992, abstract) reviews
the experimental literature on reptile learning in general and concludes
that "reptiles appear to learn what stimulus to respond to rather than how
to respond to a particular stimulus."

   An important dimension of mammalian behavior, as we shall soon see,
involves various kinds of executive processes. What about reptiles? Studies with several species of lizards (mostly skinks) have found them capable of reversal learning: they first learn to go to one, but not the other,
of a pair of stimuli for a reward, and then the experimenter reverses the
rewarded choice. Many researchers believe that success in this task requires
the executive function of inhibition (of the previously rewarded behavior).
Some lizards are skillful at reversal learning and so, by inference, at inhibition (Szabo, Noble, Byrne, et al., 2019; Szabo & Whiting, 2020), which
represents a form of behavioral control that operates after decision-making,
during action execution. But in a different study, lizards failed in a reversal-
learning paradigm that tested set shifting (basically, given a stimulus set of
black and white squares and circles, the animal has to first reverse from,

e.g., black to white, and then later shift dimensions altogether and focus on discriminating squares versus circles; Szabo et al., 2018). The issue is that the lizards did not seem to be bothered by such a shift, as mammals are, thus indicating that the original reversal learning was not based on what executive-function researchers call an attentional set. The lizards were not learning a categorization of stimuli by which to guide their future decision-making but rather learning something more concrete and limited.

Another task often used to measure inhibition is the detour task. In this task, as used with lizards (Szabo, Noble, & Whiting, 2019; Szabo & Whiting, 2020), individuals first learn that they can obtain food from inside an opaque tube by going around to the end opening and entering. Food is then placed in a similar tube that is transparent. Individuals' natural tendency is to go directly for the seen food. But some lizards inhibit that response in favor of going around to the end opening and entering (as they had learned previously for the opaque tube), some on their first trial. This would seem to indicate that at least some lizards can inhibit prepotent behavioral responses. (However, it is possible that the lizards just generalized to the transparent tube what they had learned with the opaque tube previously, i.e., going around to the end opening.)

Lizards' ability to inhibit a natural or learned behavior in these ways provides further evidence that they are indeed making decisions in a way that *C. elegans* is not. Nevertheless, inhibition is only one limited skill from the suite of skills associated with executive control in mammals; indeed, in this case it is the simplest form, sometimes called "global inhibition" or a "stopping mechanism," which is associated mainly with go-no-go decisions in which performing the wrong action might be more costly than inaction (Aron et al., 2014). Global inhibition is likely of special importance when, for example, an individual is in the process of eating and a predator approaches. The individual must then "freeze" its eating behavior. Then, separately, it decides what to do in the new situation (e.g., flee). Global inhibition thus enables goal-directed agents to operate with go-no-go decision-making sequentially across different actions, turning one off and another on as appropriate to the goal and situation. (Discrimination learning among alternatives is only possible for such creatures if one alternative is above threshold and the other is below.) This is opposed to an either-or process of decision-making in which the individual simultaneously considers more than one behavioral option simultaneously (which mammals arguably do; see chapter 4). The conclusion is thus that lizards make decisions and can inhibit the execution of bad ones, in effect taking a response that was "go" and changing it, in medias res, to "no-go."

My hypothesis is that the basic process of pursuing goals and making go-no-go decisions is the same for the vast majority of flexibly behaving animal species, both extinct and extant, including fish, reptiles, amphibians, and even some invertebrates such as bees and spiders, who may have evolved these capacities in parallel. What differentiates species with respect to this dimension is which of their behaviors are goal-directed and decision based, in what degrees, with many complex behaviors, such as lizards' foraging behavior, having some components that are more stimulus driven and others that are more goal-directed and decision based. Thus experiments show that garter snakes' "attack" behavior against certain objects is strongly and inflexibly driven by certain stimuli (Burghardt, 1966), whereas their search for prey leading up to the attack is more flexibly goal-directed and decision based. The conclusion is therefore that reptiles and many other organisms operate similarly as basic feedback control systems with the same basic structure of goal pursuit, go-no-go decisions (with global inhibition), and discrimination learning. They are operating as goal-directed agents.

Ecological and Experiential Niches

The process of evolution by means of natural selection builds organisms that are capable of producing effective actions. This requires perception of the environment, but not everything in the environment. Organisms need to perceive only those aspects of the environment that are relevant for their actions. A thermostat senses only temperature because that is all it needs to perceive to do its job. *C. elegans* perceives nutritious and noxious chemicals because that is all it needs to perceive to obtain food. A lizard perceives many things because that is what it needs to perceive to direct and control its various effective actions. The organism's action capabilities thus determine its experiential world. (See J. J. Gibson's 1977 argument that an organism's perceptual world comprises "affordances" for its actions.)

The straightforward implication is that organisms that behave differently experience the world differently. In his (in)famous book *A Foray into the Worlds of Animals and Humans*, Jakob von Uexküll provides a charming and somewhat radical description of the situation (1934, p. 43):

We begin our stroll on a sunny day before a flowering meadow in which insects buzz and butterflies flutter, and we make a bubble around each of the animals living in the meadow. The bubble represents each animal's environment [Umwelt] and contains all the features accessible to the subject. As soon as we enter into one such bubble the previous surroundings of the subject are completely reconfigured. Many qualities of the colorful meadow vanish completely, others lose their coherence with one another, and new connections are created. A new world arises in each bubble.

Each species interacts with the environment in its own unique way—from worms to butterflies to bees to bats to octopuses to sponges—which means that, in an important sense, they are not living in the same environment at all; they live in different worlds, different bubbles. This is the situation that biologists attempt to capture by saying that each species lives in its own ecological niche, which, from the point of view of the organism's perception, means in its own experiential niche. A worm's experiential niche is basically dirt and bacteria; for the worm, fish and trees and humans simply do not exist, because it has no truck with them. A lizard's experiential niche, in contrast, comprises ants, crickets, grass, a burrow, and many other things—visually and auditorially registered—because that is what the lizard needs to perceive in its ecological niche to support its foraging and other activities.

Thus, in an important sense, organism and environment codetermine each other (see George Herbert Mead's epigraph to this chapter). But they do so in different ways. We often speak of individual organisms as being "adapted" to the environment, but this does not mean that the individual has been shaped by it. Indeed, evolutionarily, we should not say that the individual has adapted at all; it is the species (i.e., the population) that has adapted, changing in its population-level characteristics over time, as those of its individuals who did not have what it takes to survive and reproduce have been eliminated. But natural selection does not cause the individual to be a certain way; that is the job of (epi)genetic expression during ontogeny, and it happens before any selection takes place. Indeed, Darwin's genius was precisely in seeing clearly that individual variability occurs ahead of time and independently of the selection process. Natural selection thus operates like a giant sieve, with each hole in the sieve allowing myriad different shapes or types of individuals to pass through, so long as they are "small" enough. The sieve eliminates individuals who are incompetent; it does not shape the competencies of those who survive. And so we may say that the organism and its actions determine both its behavioral niche and its experiential niche.

In the case of agentive organisms, the individual's goals and actions determine its experiential niche in an even more radical way. An organism's agentive actions in the moment are guided not by its perception but by its *attention*. That is, as an organism is preparing for goal-directed action, it may perceive all kinds of things, but to make an effective decision, it must *attend* to some subset of these perceptions, namely, the subset that is *relevant* for its goal pursuit. An organism that is 100 percent stimulus driven has no goal, so there is no such thing as selectively attending to relevant things; without goals there is nothing to be relevant to. It is thus unlikely that unicellular organisms, as completely stimulus driven, use attentional processes to sense relevant situations. But lizards and other goal-directed agents pursue goals and make behavioral decisions, so they must attend to those aspects of their perceived environment that are relevant for those goals and decisions. Attention is thus a kind of goal-directed perception.[1]

For behaving agents, what is relevant for attention is not objects but situations. The reason is that the individual's goals and reference values are represented as fact-like situations to be pursued. Although we sometimes speak of an object or location as a goal, in reality the goal is *having* the object or *being at* the location (Davidson, 2001). So if goals and reference values are internally represented as desired situations, then to produce an effective goal-directed action, the organism must attend to relevant situations in the environment using a similar representational format. Imagine a waiting lizard with the goal of ingesting a cricket. A cricket approaches. To determine what to do, the lizard attends to several situations (or facts) immanent in a single perception: the cricket is the right size for easy capture; the cricket is high in the bush; the cricket is moving closer; and so on. These different situations are not different perceptions; they are all present simultaneously in the same perceptual image on the lizard's retina. These are the *relevant situations* to which the lizard must *attend* if it is to make a good decision about whether to pursue its goal of eating a cricket, given that going after the cricket requires various actions. Relevance is thus determined by the organism's goal, as the lodestar of its behavioral decision-making.[2]

Along with goal-directed agency, then, comes a fundamental shift in experiential niche. Organisms no longer just perceive attractive and repulsive stimuli; they attend to situations that are relevant for their goal pursuit. Situations that are relevant for their goal pursuit are of two types: (i) opportunities for goal attainment (e.g., the cricket is low in the bush); or (ii) obstacles to goal attainment (e.g., a snake is close by). Opportunities and obstacles are defined, obviously, in terms of the organism's action capabilities. Goal-relevant opportunities and obstacles for action constitute a completely new type of experiential niche: the agentive niche. Although changes in a species' experiential niche most often result from changes in specific actions—for example, changes in foraging behavior for a new kind of prey lead to new perceptual capabilities with respect to that prey—in this case, a change in the basic organizational structure of the individual's action production leads to a change in the basic structure of its experiential niche. Anthropomorphizing the process of evolution by means of natural selection, we may say that to empower organisms to deal effectively with unpredictably changing environments, Nature devised a new way of operating, agency, in which the individual directs its actions flexibly toward goal situations and controls its behavior flexibly via attentionally informed decision-making, which requires it to attend to goal-relevant situations in the environment as either obstacles or opportunities for goal attainment. This is an entirely new *type* of experiential niche, one that nonagents simply do not experience.

Figure 3.4 depicts the organization of the basic feedback control system for goal-directed agents (at one level of the hierarchy). In this diagram, the thermostat's simple perception is replaced by attention to relevant situations, and the thermostat's mechanical comparator is replaced by behavioral decisions instantiated as go-no-go decisions. Inhibition is conceptualized here as simply the no-go option (e.g., if a lizard is feeding and a predator appears, the first thing the lizard does is stop feeding). The organism attends to the effects of its actions (light-gray upward arrow) as both immediate feedback for possible adjustments and for purposes of longer-term learning.

Foundations of Agency

Behaviorism has died, but its legacy persists in behavioral scientists' tendency to speak in terms of stimuli and responses, as if organisms were sitting around waiting to be prodded into action. This may be true of amoebas

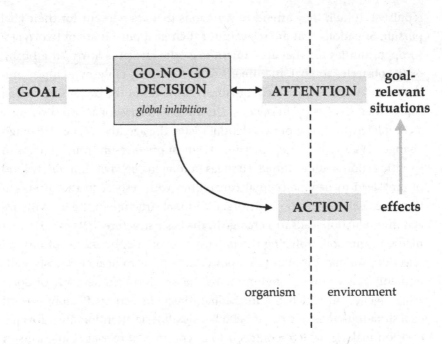

Figure 3.4
The organization of feedback control systems for goal-directed agents.

and other stimulus-driven organisms, but it is not true of goal-directed, decision-making agents. Goal-directed, decision-making agents actively seek to fulfill their goals and maintain their reference values by acting on the world; they are doing this basically constantly, even when they are waiting in front of an experimental apparatus for the appearance of food. And goals are not mysterious entities, as behaviorists would have it, but simply off-line perceptions of the world—perceptually imagined situations—that the organism desires or is motivated to bring about. It then behaves until it perceives the realization of those desired situations in the actual world (Powers, 1973).

The foundation stone of behavioral agency is thus feedback control organization, as found in lizards and other goal-directed agents. Goal-directed agents are not just stimulus driven. They direct their actions toward goals, a process that is the sine qua non of intelligent action, because without a goal there can be no sense of effectiveness or success. Goal-directed agents also control their actions via informed decision-making, accompanied by the possibility of inhibiting unwanted actions, leading to new forms of behavioral flexibility. Such informed decision-making requires the individual to attend to situations that are relevant for particular goals as either opportunities or obstacles. And this may be the key difference from machines, such as our leaf vacuum machine, which only sense the environment in hardwired ways, without the ability to flexibly and selectively attend to situations that are relevant to their goals.

Nevertheless, despite functioning as flexible decision-makers, goal-directed agents can make only simple decisions. They do not survey and choose among multiple behavioral possibilities simultaneously but rather move sequentially from one go-no-go decision to the next. This is to be expected of an organism whose behavior emanates exclusively from the single psychological tier of perception and action, rather than from, in addition, an executive tier of decision-making and cognitive control that formulates multiple action plans and then decides among them before acting, as do more complex agents. If we followed different evolutionary paths from urbilaterians, we would very likely find other goal-directed agents based on processes of parallel evolution. In particular, the flexible behavior of eusocial insects such as ants and bees suggests that they too—at least in some components of some of their behaviors—are pursuing goals and making decisions. Thus, when ants are foraging for food, experimental data suggest that they take a kind of visual snapshot of their nest as they venture away from it, and this snapshot serves as a goal destination as they make their return journey (Möller, 2012). Nevertheless, like all goal-directed agents, ants' behavior is still often stimulus driven, their decision-making is still only of the go-no-go variety, and their learning is still just learning when to perform their existing actions.

The selection pressures leading to behavioral agency are not of the specific variety. That is, they are not of the type that led Darwin's finches to evolve different foraging activities to exploit one or another particular resource. Rather, the selection pressures leading to behavioral agency arise when a population of organisms moves into an ecological niche rife with unpredictabilities, for example, clever prey and predator species. The point is that when the environment always repeats itself (e.g., in filter feeders, food is always just an open mouth and a few wiggles away), then hardwired, open-loop, stimulus-driven behavioral organization works just fine. But when novelty and unpredictability arise, stimulus-response organization leads to failure, as the individual is always "fighting the last war." With feedback control organization, Nature can still hardwire the most important goals but at the same time empower the individual to pursue them flexibly by attending to relevant situations and making informed behavioral decisions. The lizard does not choose to crave tasty crickets; it chooses how to pursue this cricket now.

This way of doing things presupposes a hierarchy of feedback control systems. Organisms' highest-level goals include things such as ingesting nutrition, escaping predators, and mating. Below that level are means for effecting those goals, including locomotion and a variety of other specific actions that can be incorporated into larger activities (i.e., serve as modules) as needed. When we view these larger activities, it may be that the full hierarchy and sequence of actions for any particular activity (e.g., lizards foraging for ants, as in fig. 3.3) have components and levels that are differentially structured in how much they are hardwired (stimulus driven) or individually controlled (goal-directed). The fact that a given activity may comprise such a complex mix of components means that judgments of whether the activity is "innate" or "learned" are decidedly unhelpful. Moreover, in many cases, the same action is incorporated into multiple different higher-level activities (as module); for example, a lizard's ability to locomote to desired locations is used in foraging, fleeing, finding its burrow, and so on. This is a different kind of modularity from the standard type in evolutionary psychology, which only considers the more molecular act. A good name for it might be "hierarchical modularity" (a more descriptive term than the "Baldwin effect").

Much evolutionary change occurs when Nature makes a change at the top of a hierarchically structured activity, which then drives the lower-level behavioral adjustments, and ultimately the biological adaptations, needed to carry it off. Thus, if a population of lizards adapts to a change in available prey by acquiring a taste for a new and different species of insect, this puts pressure on the perceptual and motor abilities necessary for hunting and capturing this new prey to adjust in kind (e.g., maybe now one needs to climb trees to capture available insects). This "trickle-down" way of looking at things provides a more nuanced perspective for addressing the question of whether a particular change in the way an organism functions is the product of many small, specific adaptations or of one larger, more encompassing adaptation. Often it is both. Thus a new ability to climb trees to capture a new insect may be considered a separate module, but the larger behavioral context within which such climbing abilities emerged—the goal of capturing the new insects—gives "direction" to the process. Neglecting the hierarchical structure of behavioral and psychological organization leads to overly simplistic views of the evolutionary process. Arguably, hierarchical modularity and trickle-down evolution provide a more accurate picture of the complex ways in which the behaviors and behavioral organization of species change over time.

In general, this hierarchical, trickle-down view of behavioral evolution suggests that agentive action is not just an object of natural selection but also a causal force in the process of evolutionary change. (Jean Piaget [1976] calls behavior the "moteur de l'evolution.") If a new insect species suddenly appears, lizards cannot be behaviorally adapted for capturing it precisely because it is new; it has never before been a part of their ecological niche. Nevertheless some individuals may be able to use their flexible, agentive powers of goal-directed attention and decision-making to extend their existing behavioral skills to make the new insect a part of their diet. Because this new extension of behavior comes into existence not because of any genetic changes but because of the exercise of behavioral agency, we may say that to some significant degree, the agent and its flexibly organized skills play a causal role in the process of evolutionary change. The organism's newly effective actions now make possible genetic changes that serve to support the organism's pursuit of the new insect (so-called genetic assimilation).

4 Ancient Mammals as Intentional Agents

The pursuance of future ends and the choice of means for their attainment are . . . the mark and criterion of the presence of mentality in phenomena.

—William James, *The Principles of Psychology*

Lizards and other reptiles are thus behaviorally flexible decision-makers—compared to *C. elegans*. But compared to mammals, their behavior is still a bit stereotyped and inflexible. The explanation is that along with the emergence of mammals some 200 million years ago came a huge new jump in behavioral agency based on a fundamental reorganization of how individuals direct and control their actions. Mammals direct their actions toward goals not just flexibly but intentionally, as they cognitively simulate possible action plans toward their goal before actually acting. And they control their behavior not just by making go-no-go decisions but also by making either-or behavioral choices as they evaluate the possible plans' likely outcomes and then cognitively monitor and control behavioral execution as it unfolds.

The evolutionarily new psychological organization that enables this new mode of functioning—what I call intentional agency—is made possible, first, by more flexible forms of motivation: individual mammals act not just toward fixed goals but on the basis of more flexible emotions and motivations that can be overridden as needed. In addition, individual mammals undergo a slowly unfolding ontogeny (life history) involving much learning and cognitive development after their emergence into the external environment, which enables individuals to unlearn and relearn things as needed. But the real power of these psychological novelties is that they enable and then also participate in a new type of psychological organization comprising not just an operational tier of perception and action—as already in goal-directed agents—but also an executive tier of decision-making and cognitive control. This new two-tiered way of operating enables individuals not only to do things flexibly but also, in some sense, to know what they are doing.

Once again I will not be considering species that branched off from reptiles in nonhuman directions, some of whom display great intelligence and agency. Of most importance are birds of the corvid and psittacine families (crows, jays, parrots, etc.). They have almost certainly evolved skills of executive control similar to those in mammals, presumably in a process of parallel evolution, but that is a story for someone else to tell.

Emotions, Cognition, and Learning

The first mammals in the lineage on the way to humans were small, squirrel-like creatures about 200 million years ago. Modern-day squirrels, which we may use as a model, are obviously different from this creature, possessing some adaptive specializations that the ancient mammals did not have, such as caching nuts for future use. Other mammals, both ancient and extant, have their own adaptive specializations. But what we are concerned with here is the organization of mammals' most basic modes of psychological functioning—not what things they do but how they do them—and that, by hypothesis, is generally similar across mammalian species.

The ecological conditions that led early mammals to their new ways of doing things involved, once again, unpredictabilities in their ecological niche. The key for early mammals was their new social niche. Whereas reptiles are mostly solitary foragers, most mammals live and forage in a social group of one type or another. Thus, in addition to complexities created by their foraging ecologies, early mammals had to deal with complexities created by enhanced competition with group mates for food and other resources. When the whole group finds a patch of food at the same time, there is a premium on making fast, efficient foraging decisions. Since those group mates are in exactly the same situation, a kind of "arms race" of cognitive skills for outcompeting conspecifics for access to food can occur. (Imagine, once again, our leaf vacuum machine, but this time competing with other similar machines to get the most and best leaves.) In addition,

Figure 4.1
Imagined early mammal 200 million years ago.

many mammals compete with others by cooperating with coalition partners in teams based on social relationships. Operating in this newly complex
socioecology led early mammals to evolve a new manner of functioning
involving not only new psychological skills and motivations but also new
ways for making decisions more efficiently.

Early mammals' new manner of functioning comprised three new psychological capacities, organized in a new way. First, mammals evolved new
ways to motivate their action: not just by an ineluctable attraction to a
goal, but rather by more flexible motivations and emotions. Instead of a
more or less fixed reaction to a predator, for example, mammals evolved
an internal psychological state, the emotion of fear, that precipitated an
"action tendency" (Frijda, 1986) to flee, but with the flexibility to do something else if that would be more beneficial. Or if a conspecific attacked it, an
individual mammal would not just retaliate straightaway but would experience the emotion of anger, which would precipitate an action tendency to
fight, with exactly how or whether to fight being adjusted to the perceived

value of other behavioral options. The point is that emotions and motivations produce different strengths of "action tendencies" for specific actions, and these can then compete for execution. Mammalian decision-making thus became more flexible and at the same time more complex: individuals make either-or choices among simultaneously motivated alternatives. In terms of brain bases for these new motivational mechanisms, classic views attribute to reptiles a completely nonemotional reptilian brain that lacks a limbic system, which contrasts with the emotional brain of mammals (P. MacLean, 1990). Modern research now downplays the differences between reptilian and mammalian brains (e.g., Naumann et al., 2015), but it is still the case that the "limbic system" (however that is now conceptualized) seems to play a more important role in mammalian than in reptilian behavior.

Second, mammals evolved a number of new and more flexible cognitive capacities for making quicker and better decisions in the context of heightened social competition. Thus social competition with group mates puts a premium on the quick and accurate assessment of how best to take advantage of a particular foraging opportunity (different for different species), and to avoid obstacles while doing so. This might involve everything from sophisticated skills of spatial or temporal cognition to sophisticated skills for predicting the behavior of conspecifics. Further, social competition with group mates puts a premium on avoiding costly mistakes in decision-making, effected, for example, by planning before acting via imaginative cognitive simulations and then monitoring and supervising execution of the planned action. In terms of brain bases for these new cognitive skills, mammals have a six-layered neocortex that greatly exceeds the three-layered structure of reptiles in number of neurons and neural subtypes, as well as many more functionally differentiated areas and a corpus callosum between hemispheres that facilitates more rapid information processing (Molnár, 2011; Kaas, 2013).

Third, mammals evolved a new ontogenetic (life history) pattern involving much more learning. Individual mammals develop slowly and in interaction with the environment (rather than quickly and inside an egg, as in reptiles). In the early stages of life, mammalian infants can count on their mother's milk for nutrition and on her vigilance for their protection and safety. This life history pattern is costly for mothers and risky for infants. A major compensating advantage is that the infant, who does not have to spend time and energy seeking food and escaping predators, can focus on learning about its local environment (and learning in play, which is arguably absent in reptiles). By the time they reach adulthood, individual mammals have learned many things, which can be unlearned or modified more easily than more hardwired behavioral adaptations. Mammals also engage in a new type of learning: instrumental learning (not to be confused with instrumental or operant conditioning) based on a new understanding of how their own actions causally affect outcomes in the environment. Learning and developing in interaction with one's physical and social environments—and unlearning as needed—add immensely to the behavioral flexibility of mammals.

The way that mammals direct their actions via more flexibly motivated goals and more powerful skills of cognition and learning leads to a more flexible version of our basic feedback control system. The key to this new version is the functioning of a wholly new tier of psychological organization: the executive tier. This new executive tier enables more flexible forms of planning and decision-making that output not an action but an intention to act, with the further possibility of self-regulating the fidelity of the intention's translation into action. Mammals thus became intentional agents.

The Executive Tier

The study of executive function—also referred to as cognitive control—has classically focused on individual differences in humans, especially those with brain damage or with cognitive deficits due to aging. The field is rife with interesting phenomena and confusing terminology. A representative definition of the general phenomenon goes as follows (Banich, 2009, p. 89):

> Executive function is a process used to effortfully guide behavior toward a goal, especially in nonroutine situations. Various functions or abilities are thought to fall under the rubric of executive function. These include prioritizing and sequencing behavior, inhibiting familiar or stereotyped behaviors, creating and maintaining an idea of what task or information is most relevant for current purposes (often referred to as an attentional or mental set), providing resistance to information that is distracting or task irrelevant, switching between task goals, utilizing relevant information in support of decision-making, categorizing or otherwise abstracting common elements across items, and handling novel information or situations.

The reason that pretty much all definitions include a list (this one contains eight items) is that the field identifies strongly with the clinical tasks used to assess executive function, and they are many and various. The tasks are important because performance on them predicts all kinds of important things about the psychological functioning of individual human beings. The problem from the current point of view, however, is that the vast majority of tasks focus on humans and their ability to follow and stick to abstract, linguistically expressed rules proposed by experimenters (such as in the Stroop task, the Wisconsin Card Sorting Test, the digit span memory task, and many others governed by explicit rules). Following explicit rules generated by others is not a skill that squirrels or other nonhuman mammals need, and so arguably the core phenomena of executive function comprise more basic processes related to self-generated goal pursuit.

Cognitive scientists have made various attempts to bring order to such lists, but a widely accepted typology comprises (i) inhibition (including inhibitory control, self-control, and behavioral inhibition, as well as interference control, selective attention, and cognitive inhibition); (ii) working memory (i.e., holding information in mind and mentally working with it in various ways); and (iii) cognitive flexibility (e.g., set shifting, mental flexibility, or mental set shifting, closely linked to creativity) (paraphrased from Diamond, 2013). This formulation is helpful, but the terminology is still confusing in that "inhibition" would seem to be a basic psychological process, "working memory" would seem to be a cognitive mechanism within which basic processes work, and "cognitive flexibility" is a trait that people and processes possess. So what we need for current purposes is a more coherent formulation. I thus propose here a separate executive tier of psychological monitoring and control, which is itself a feedback control system with new forms of decision-making and behavior monitoring that facilitate the organism in directing and controlling its actions.

The most basic distinction needed is between reactive and proactive forms of executive functioning. According to Braver (2012, p. 106):

The proactive control mode can be conceptualized as a form of "early selection" in which goal-relevant information is actively maintained in a sustained manner, before the occurrence of cognitively demanding events, to optimally bias attention, perception and action systems in a goal-driven manner. By contrast, in reactive control, attention is recruited as a "late correction" mechanism that is mobilized only as needed, in a just-in-time manner, such as after a high interference event is detected. Thus, proactive control relies upon the anticipation and prevention of interference before it occurs, whereas reactive control relies upon the detection and resolution of interference after its onset.

This basic distinction draws a potential dividing line between reptiles and mammals. Recall that the experimental evidence for executive function in lizards was their success in reversal learning and a single detour learning task. My claim is that in these tasks the lizards were employing only reactive control, activated only in reaction to changed contingencies or a perceived obstacle, in which case they inhibited their previously successful response. This would not seem to require two separate tiers of psychological functioning—and certainly not proactive processes of executive control—but only a simple process of global inhibition that enables goal-directed agents to "freeze" an ongoing action if another more urgent situation arises.

In contrast, mammals engage in proactive executive control, including the prevision of error. In contrast to simple reactive inhibition, this manner of operating requires two separate tiers of functioning: the operational tier of perception and action plus an executive tier. The executive tier oversees the operational tier, as it were, and attempts to facilitate behavioral decisions via action planning and cognitive control. This enables individuals to proactively form and pursue plans while also reacting to unanticipated obstacles or intrusions along the way, as needed. This more deliberate and flexible way of doing things is what is often referred to as intentional action (Bratman, 1987). It requires individuals to cognitively simulate in an organized way their own potential actions, the potential obstacles and opportunities for those actions, and the probable outcomes of those actions. They do this by perceptually imagining all these action elements in the common cognitive workspace and representational format provided by an executive tier of operation. This new form of mental activity is supported by mammals' greatly expanded prefrontal cortex—the widely recognized home of most executive functions—as compared with reptiles and other goal-directed agents (Molnár, 2011; Kaas, 2013).

Mammals' actions are thus not just goal-directed but intentional; cognitive simulation and planning enable the individual to organize and choose its actions more flexibly and on their relative merits. Bruner (1973) argues that the essence of intentional action is that the individual has available ahead of time multiple possible actions to the same goal, which it can try out as needed until it achieves goal success. Piaget (1952) emphasizes that in intentional action the individual has the goal "in mind" ahead of time throughout planning and execution—that is, on the executive tier—as it decides what to do and does it. These complexities are most clearly evident when the organism's actions must be sequenced in a particular way to achieve a goal, or else one action must be embedded within another as a subplan for doing things like removing an obstacle on the way to the goal. Thus, as a squirrel plans a trek along a tree branch to fetch a nut, the squirrel might not execute that plan until it can first figure out a subplan for removing a dead branch that is blocking the way. (Some researchers would emphasize that such processes require "working memory.") How flexibly and effectively an organism can accomplish such planned action is determined both by the species' particular cognitive skills and by the individual's particular learning experiences.

The main difference between the goal-directed agency of reptiles and the intentional agency of mammals is thus that mammals not only operate on an operational tier of perception and action but also supervise this operational tier from an executive tier of decision-making and control. In figure 4.2, the main shaded components of (i) goal (now as goal/motive), (ii) attention (to relevant situations), and (iii) action (and its effects) are the same or analogous to those of goal-directed agents, as these three components form the foundation for all feedback control systems. For example, the system still contains decision-making, just in a new form. What is completely new is the executive tier of decision-making and cognitive control at the top of the diagram, in italics, which operates as its own feedback control system. The executive goal is to make a "better" behavioral decision than would otherwise be made by the operational tier. The individual attempts to make a better decision—in the context of its cognition, knowledge, and values—by imaginatively simulating multiple action plans and their anticipated outcomes proactively, and then evaluating those plans and their outcomes to make an either-or decision. This results not in an action but in an intention to act, which guides the organism's action by serving as a kind of template at which action execution aims. This process constitutes a kind of cognitive control, as it serves to keep things on track throughout the performance of an intended act. As a part of this process, the reactive inhibition of actions by reptiles transforms into a proactive inhibitory control, enabling individuals to inhibit action execution more

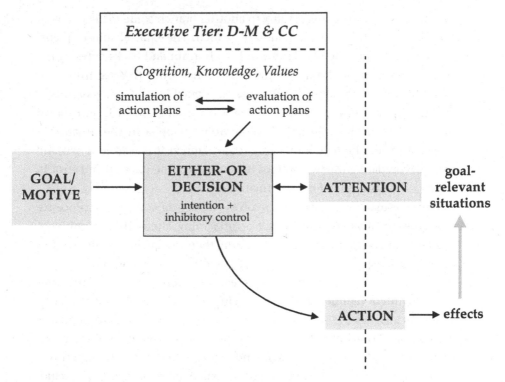

Figure 4.2
The organization of mammals' intentional agency. Shaded components are analogous to, or the same as those for, reptiles and other goal-directed agents (with some internal changes, e.g., a new type of decision-making). Italicized components at the top represent mammals' unique tier of executive decision-making (D-M) and cognitive control (CC).

flexibly on the basis of *anticipated* outcomes and in comparison to alternative behavioral possibilities. In the next two sections, following their natural ordering, I treat in more detail executive decision-making and then executive (cognitive) control.

Executive Decision-Making

In contrast to lizards and other reptiles, squirrels and other mammals often make either-or decisions prospectively among cognitively represented behavioral options before they act. Consider the following observation. A squirrel is perched on a tree branch and is trying to decide whether to get to a new location by either leaping to another branch some meters away or climbing down the trunk and out that branch. The squirrel coils in preparation to jump but then backs down. It coils again and backs down again. Finally it gives up and ambles down the branch onto the tree trunk and then out the desired branch. What has happened here? One possibility is that the squirrel has run a kind of cognitive simulation. It has imagined (in a kind of off-line perception) what would happen in the situation if it leaped for the branch, and what would happen if it walked down and around, comparing the two options in a process of mental trial and error in which failure is not fatal but informative.[1]

Decision-making of this kind involves two sets of component processes in a constant dialogue: simulating or imagining alternatives cognitively, and evaluating each so as to choose among them. To begin, cognitive simulations obviously require some form of cognitive representation, as well as the ability to manipulate those representations imaginatively. For mammals, cognitive representations are exclusively perception based, that is, iconic or imagistic. (This does not mean exact replicas of perceptions, as the representations may be categorical or image schematic.) These representations are then used to imagine nonactual situations (although this is likely constrained to types of situations experienced before, thus excluding counterfactual representations). The content of these representations is potential situations in the environment in combination with potential acts and their potential consequences in those situations. The fact that attention to situations in the environment is mainly based on vision, audition, and other senses, while action plans are based on proprioception, is a main reason why an executive tier is needed. Such cross-domain comparisons require a common workspace and representational format in which intentional actions and environmental outcomes may be imagined together (presumably in a perception-based, image-schematic format). The precise experiential content of these representations depends, of course, on the cognitive capacities and experiential niche of the species.

The process of cognitively simulating possible actions and their outcomes is called planning. Lacking an executive tier, lizards and other reptiles do not engage in planning their actions, whereas squirrels and other mammals do. This difference might thus reflect the distinction between more goal-driven or emotion-based action tendencies (over which Nature keeps more control, typically because the challenge is urgent, so that fast responses are required) and those involving one or another form of thinking or action planning (which require more time to simulate and choose among possible action plans) (Kahneman, 2011). In this cognitive way of doing things, simulating action plans and their effectiveness involves one or another form of "predictive processing" in which the individual anticipates (again by imagining) what the environment will look like as a result of realizing various behavioral possibilities (e.g., Clark, 2015).

In evaluating the simulations, researchers in animal behavior and cognition assume (borrowing from human decision science) that choosing among multiple action plans involves consideration of (i) the values of the expected outcomes of the different action plans, and (ii) the probability that those different action plans will actually achieve those expected outcomes (based on a cognitive assessment of the relevant opportunities and obstacles at hand and the possible actions available). This evaluation process operates both when the organism is deciding among two possible action plans to the same end goal (in which case the only consideration is the probability of success of each plan), and when the organism has two possible action plans, each with a different expected outcome. If the evaluation of all plans is negative, then the organism will engage in another round of action planning and expected outcome evaluation—perhaps involving the complex sequencing or embedding of plans—before action. The "best" action plan, of whatever degree of complexity, may then be executed. Such dialogue between imagined action-outcomes pairs and their evaluation can only take place in an executive tier of functioning.

Three lines of experimental evidence support the proposal that squirrels and other mammals operate in this general way. First, Chow et al. (2015) gave five gray squirrels a reversal-learning task, at which they were, like lizards, quite skillful. In addition, unlike lizards, the squirrels showed evidence of simultaneously considering behavioral options before acting. Specifically, as they were in the process of learning to go from one option being rewarded to the other option being rewarded—the evaluations at that point being close to identical—they often paused and looked back and forth from one option to the other (what the authors call "head-switching"). This particular behavior might have other interpretations, but in fact a similar behavior has long been observed in squirrels' rodent cousins, rats, going all the way back to Tolman (1948), who called it "vicarious trial and error." Thus rats also show robust skills of reversal learning, and they also often pause and look at both options before choosing. Redish (2016) reviews the data and argues that indeed what the rats are doing is making a mental decision among mentally represented behavioral options, providing neurophysiological evidence to support his case. It is worth mentioning, in addition, that rats also show in reversal-learning paradigms robust skills of set shifting and extradimensional shifting (which lizards do not), suggesting the formation of cognitive sets that might be used in more complex action planning (see Szabo et al., 2018, for a review).

Second, Chow et al. (2019) confronted gray squirrels with an apparatus that enabled them to retrieve a nut either by pushing one of five levers or by pulling a different one of those five levers. Squirrels naturally preferred retrieving the nut by pushing the first lever. Then, some months later, researchers modified the apparatus so that the preferred strategy of pushing was physically blocked. All five squirrels adapted flexibly to the modified problem and solved it by pulling the other operative lever—on their first trial. Squirrels seemingly imagined the old action failing (prevision of error) and perceived which new action would succeed. This is, in effect, a detour task requiring an updating of strategies, and again squirrels' rodent cousins, rats and mice, also perform very well in detour tasks (see Kabadayi et al., 2018, for a review). And again in these detour tasks, rats and mice show a distinct pattern of pausing and alternating attention at their choices before acting, again suggesting a cognitive process of assessment and choice (Redish, 2016; Juszczak & Miller, 2016). Further, Blaser and Ginchansky (2012) report experiments showing that rats travel to multiple food locations with differently valued rewards in a semiefficient, seemingly planful manner, suggesting at each decision point an either-or choice among differently valued destinations. In a general review, Crystal (2013) presents several lines of evidence that rats cognitively represent events that they are anticipating to occur, and incorporate such representations into acts of action planning.

Third are experiments on a behavior very similar to our hypothetical squirrel deciding whether to leap to a faraway branch or walk down and around to its desired destination. These experiments have been performed not with squirrels but with rats. Both Foote and Crystal (2007) and Templer et al. (2017) presented rats with a discrimination task in which they had a choice on each trial: the rat could either solve a problem and get a large reward or, if it was anticipating failure, opt out and get a smaller reward for free. When the discrimination was easy, the rats almost always chose to solve the problem and get the larger reward. However, when the discrimination was difficult, they opted out and went for the free smaller reward (a good decision, since they often made mistakes on similar trials when not given the opt-out option). The rats anticipated failure (prevision of error) based not on an alternative solution that they could see (as in the Chow et al., 2019, apparatus with squirrels) but on their own uncertainty in making a behavioral decision. (In a similar study, Smith et al., 1995, found similar positive results in a bottlenose dolphin.) Some controversy surrounds how to interpret these studies—whether the individual knows that it does not know, or simply perceives its sense of uncertainty—but in either case their behavioral choice relies on an anticipation of error based on an executive assessment of different action possibilities. The rats were cognitively simulating the choice of an action and its likely failure, so they chose another simultaneously available action possibility instead (see Smith, 2009, for studies with other, mostly primate, species supporting this interpretation).

Different species weigh the two parameters of action evaluation—outcome value and outcome probability—differently. The relative weightings of these parameters for a species characterize its risk profile. Thus a species that goes for a high-value resource no matter the probability of success is called risk prone, whereas a species that goes only for resources that it is highly likely to acquire no matter the value (over a certain threshold) is called risk averse. For example, in experiments, chimpanzees tend to leap at every opportunity to acquire a pile of bananas no matter how unlikely they are to get them, whereas bonobos typically opt for higher-probability options with less-exciting payoffs (Heilbronner et al., 2008). And different mammalian species discount the value of delayed rewards to differing degrees, presumably based on differing degrees of risk tolerance (Stevens et al., 2005).

These species-level risk profiles are set to some degree by Nature, as it were, but much room for individual agency exists in mammalian risk-taking as well. Thus individuals of many species make individual foraging decisions that are both state sensitive and context sensitive. For example, individuals of many species are more likely to take risks when they are in a more resource-deprived state than otherwise, presumably because the safe options are judged to be insufficient to overcome the deprivation (see Kacelnik & Mouden, 2013, for a review). Individuals of many species are also sensitive to situational cues of risk based on individual learning. For example, in an experiment, rats were given a choice between pressing a lever for a small, safe reward or pressing a lever that had the possibility of a much larger reward but also the possibility of punishment, with the risk of punishment rising across trials. The rats preferred the large reward when the risk of punishment was low, but then across trials, as the risk of punishment increased, they began to prefer the safer option (Simon et al., 2009). The point is that state-sensitive and context-sensitive decision-making provides further evidence that the individuals themselves are making either-or, value-based decisions among multiple alternatives.

Individuals are, of course, not computing outcome values and outcome probabilities and combining them mathematically. More likely, they are employing what has been called ecological rationality (e.g., Todd & Gigerenzer, 2012), in which individuals work with simple heuristics either given to their species by Nature or learned individually. There may be some overarching ecological principles leading to similar heuristics across species based on factors such as group size (as a proxy for social competition; Hintze et al., 2015). There may also be species-unique heuristics based on special feeding ecologies; for example, researchers hypothesized that the difference in risk profiles between chimpanzees and bonobos noted earlier developed because chimpanzees' feeding ecologies involve much more unpredictability. And individuals can also learn risk heuristics in situations for which they could not be genetically prepared; for example, many vervet monkeys have individually learned to use the sound of cowbells to predict risk from human herders and so to stop feeding and flee (and cowbells have only existed in the monkeys' habitat for a few decades; Cheney & Seyfarth, 1991).

One potential piece of contradictory evidence is that great apes (and so, presumably, other mammals) are not capable of simultaneously representing incompatible outcome possibilities. For example, apes do not understand that a piece of food dropped into an upside-down, Y-shaped tube might possibly come out either side at the bottom (Suddendorf et al., 2017). But this is different from the current claim because it is about understanding how the external world works. My claim is that when mammals understand the external world sufficiently, they can consider simultaneously which of two actions—their own actions—is most likely to produce a desired result. Thus, in the opt-out paradigm, the rats know that they have two choices in front of them—one risky, one safe—and attempt to figure out which choice will lead to the most food. And the squirrels in the middle of a reversal-learning regime know that either option is possibly correct, and they attempt to figure out which option would be best. These behaviors do not involve conceptualizing two incompatible possibilities in the environment; rather, they involve choosing which of two actions that have been successful in the past is most likely to be successful now, given an easily understandable environmental situation.

Overall, I believe that the experimental studies with squirrels and rats support the hypothesis that they, and presumably other mammals, often consider simultaneously two cognitively simulated behavioral options in acts of either-or decision-making. The output of such executive decision-making is an intention to act. An intention to act is more than a goal; it is a plan for achieving a goal (Bratman, 1987). Goals provide direction to the organism's action planning and evaluation, but they do not translate directly into action. I can have a goal to become rich but still not be doing anything to achieve that goal at the moment; but if I intend to get rich, you may reasonably ask what I am doing, or plan to do, to achieve my goal. An intention is a plan that has been chosen or decided on toward a goal, and it then supervises, as it were, action execution to keep it on track. Although an intention is an action plan, it is still mutable in the sense that it may change because of perceptual feedback from action outcomes. On the basis of such feedback, the individual may decide either that the plan was well executed but faulty, or else that the plan was potentially good but the action execution was faulty. This second step in the process is most often called executive or cognitive control.

Executive (Cognitive) Control

After a decision has been made, and an intention to act has been formulated, the executive tier supervises behavioral execution. What this means specifically is that the executive tier cognitively monitors and controls action execution in relation to the intended action and stands ready to right the ship if necessary. If unexpected obstacles are encountered, the executive tier can cycle back to further decision-making to reevaluate the options; if the intended action is executed poorly, it can try to execute it again, only better; if unplanned actions threaten to intrude, it can act to inhibit them. Unplanned actions comprise prepotent (stimulus-driven) responses that are more or less hardwired as well as learned habitual behaviors that have been effective in similar situations previously. Thus, while in their executive decision-making squirrels and other intentional agents are attempting to anticipate and plan for obstacles and errors, in their cognitive self-monitoring and control they employ a kind of post hoc fail-safe mechanism for detecting potential problems and intrusions during the actual performance of the action.

That squirrels and rats often choose an option different from the one that has been successful in the past—in experiments, sometimes on the first trial—suggests that they are in some way predicting that the previously successful action would now fail (prevision of error), and they are inhibiting it, or at least devaluing it relative to other options, in the decision-making process. Berkman et al. (2017) argue that what is often called self-control in humans might best be thought of simply as a special application of the normal process of "value-based choice," in which the option to be avoided is devalued, and the preferred option increased in value, relative to the other. So mammals are very likely not engaging in global inhibition in the context of go-no-go decision-making, as lizards do, but rather are engaging in value-based choices in which they use various cognitive processes to devalue previously successful actions in comparison to other available options.

In an extremely large-scale study, E. MacLean et al. (2014) administered two tasks of inhibition to approximately thirty mammalian species, from dogs to squirrels to elephants to various species of primates. In the A-not-B task, animals first found a food object three times in location A and then saw it moved to location B. To search efficiently in location B, they had to inhibit their tendency to search in the previously successful location A. They were given only a single trial and had to go directly to B in that task (ignoring A) to be counted as correct. In the cylinder (detour) task, animals first extracted food through the top opening of an opaque cylinder and then saw the food placed in a transparent cylinder. To retrieve the food efficiently, the animals had to inhibit their natural tendency to go directly for the food (visible through the transparent plastic) and instead retrieve it through the top opening of the cylinder. The outcome was that all mammalian species were successful in one or both of these tasks—with a strong correlation between performance on the two tasks—thus demonstrating impressive skills of inhibitory control and an "updating" of strategies. (In the study of Bray et al., 2014, twenty-four of thirty domestic dogs were suc-cessful in the A-not-B task on their first trial.) Relevant here as well is the study by Chow et al. (2019) in which squirrels used an updating strategy to adjust their actions when the apparatus was changed.

In my assessment of skills of inhibition in lizards and other reptiles in the previous chapter, I credited them with being able to inhibit both pre-potent actions (in the detour task) and actions previously learned as suc-cessful (in reversal learning) reactively. But in the studies of squirrels and other mammals just reviewed, individuals were not just inhibiting (freez-ing) one action reactively and then looking around for what else they could do (as the lizards did). Rather, the mammals were prospectively consider-ing simultaneously two behavioral options and inhibiting one of them in acts of either-or decision-making, as evidenced especially by their pausing and visually alternating between choices before acting, and in the opt-out studies by their active assessment of their possible options before acting. What we call inhibition in mammals is therefore perhaps just an integral part of value-based choice among multiple alternative action possibilities, in which the prevision that some potential actions will fail leads to their devaluation relative to others. This is not just simple or global inhibition but a more proactive process of what we might call inhibitory control.

The ecological conditions that prompted the evolution of inhibitory and other forms of cognitive control in mammals are, of course, lost in the mists of time. But two other findings from the study by E. MacLean et al. (2014) are suggestive. One is that species' performance on these tests of inhibition correlated with brain size. Brain size reflects a number of factors, but one important one is the complexity of the problems that individuals encounter in foraging and other activities. Thus skills of inhibitory control would be useful for an individual to delay going for a small amount of resources here and now in favor of traveling farther for a larger amount later (delay of gratification), which many mammals have been observed to do. Second, species' performance on these tests of inhibition also correlated with dietary breadth, an important dimension of complexity in making foraging decisions, as one might, for example, need to inhibit foraging for certain previously rewarding resources when more lucrative possibilities become available.

But solving difficult foraging problems efficiently would seem to be good for all species, including lizards. So why have only some species, for example, mammals, developed sophisticated skills of executive decision-making and inhibitory control? The answer, I believe, lies in mammals' new form of social complexity. Reptiles interact with other animate creatures mainly as predators and prey, and apparently their flexible skills of goal-directed agency are sufficient to deal with the ensuing uncertainties. But mammals live in social groups with conspecifics, and this means at least two things. First, individuals must engage in their normal foraging activities especially efficiently, because if they do not, their group mates will scramble to all the resources first (scramble competition). This may involve inhibiting pursuit of the closest resource if a competitor is closer to it. Second, individuals sometimes compete over resources with others directly (contest competition), so they must figure out whether and how to fight, with signals of dominance helping the decision. If they are competing with a stronger individual for a resource, they must inhibit their desire to go directly for it. These observations may explain why, among both birds and insects, the most social species seem to possess the most complex cognitive skills, perhaps including inhibitory control (though this feature has been little studied in these species; Boucherie et al., 2019; E. Wilson, 2012).

Two empirical studies support this special version of the social complexity hypothesis. First, Johnson-Ulrich and Holekamp (2020) gave a version of the detour (cylinder) task to five clans of spotted hyenas that varied in size and demographic makeup. The study found that hyenas living in larger groups—specifically, individuals who grew up in larger cohorts as juveniles—had greater skills of inhibitory control. In addition, low-ranking hyenas, who must frequently inhibit both feeding and aggression in the presence of higher-ranking hyenas, had the strongest inhibitory skills of all. (See E. MacLean et al., 2013, for potentially contradictory findings with lemurs, although lemurs all share a common ancestor not so long ago and thus are all highly related.) Second, Amici et al. (2008) investigated seven different primate species with different levels of social complexity, as indicated by the degree to which their social organization reflected a fission-fusion dynamic (in which the larger group splinters into smaller parties, which then reassemble into different parties throughout the day). The researchers hypothesized that this dynamic would require individuals especially often to exhibit inhibitory control in foraging and other activities. Supporting this hypothesis, they found that fission-fusion dynamics were positively associated with enhanced inhibitory skills (as measured by five different tests of inhibitory control), even more so than phylogenetic relations or complexity of feeding ecology.

I thus conclude that whereas reptiles may have some simple skills of reactive inhibition exercised in go-no-go decision-making, mammals exhibit more sophisticated skills of proactive inhibitory control in either-or decision-making. The most likely hypothesis is that mammals evolved these skills as instruments of cognitive monitoring and control to keep the execution of their plans and intentions on track, which had special importance in the context of complex social groups that created social competition of various kinds. In general, this more cognitively complex and controlled way of doing things is why the behavior of lions and tigers and bears seems so much less stereotyped, and so much more flexible and controlled, than that of lizards and turtles and snakes.

Instrumental Learning

If we view learning as an integral part of the organism's behavioral interactions with the world, the emergence of an executive tier of functioning should create new forms of learning, and it does. The most basic kind of learning is signal learning, including classical conditioning and discrimination learning, in which the individual learns when to perform its consummatory and other actions. In the language of behaviorists, signal learning brings the animal's already existing response under the control of a new stimulus. Virtually all vertebrate species engage in this type of learning on the perception-action (operational) tier of functioning. But the addition of an executive tier of functioning—with its proactive forms of cognitive simulation, action planning, and cognitive control—brings with it the possibility of learning not just when to perform an action, but which actions organized in which ways lead to which kinds of results in which kinds of situations. This can mean determining in the immediate situation what is working and what is not, with perceptual feedback enabling the individual to revise its plan in medias res, or it can mean learning things for the longer-term future.

The point is that when an organism attends to its own perception-action functioning from an executive tier, it can see how its actions lead to outcomes in the environment in a way that can be saved for future use. Dickinson (2001) has thus argued that rats in experiments do not just associate their act with an outcome but understand a causal relation between act and outcome. Thus, if a reward for a rat's act comes a few seconds after the act (too late to be a causal effect), the rat does not perceive the contingency.[2] Further, if a rat gets rewarded for an act but at the same time gets rewarded randomly, it does not learn the act because the random rewards undermine the causal contingency of act and effect. The conclusion, based also on other evidence, is that rats in some sense understand that their act causes the reward as effect. And this occurs in an experimental situation in which the connection between act and result is opaque (e.g., pushing a bar leads to food dropping into the cage); even more natural should be the connection between an action like pushing an object and its moving, or an action like biting an object and its breaking apart. And so the claim is that mammals have evolved to perceive the way that their acts cause effects in the environment and to learn that connection.

This kind of instrumental learning based on a causal analysis of actions and their effects—as opposed to classical conditioning or discrimination learning—requires an executive tier of functioning. Specifically, to learn instrumentally, the organism must experience its own actions and their effects—with reference to the intended act and its intended effects—from an executive tier of cognitive monitoring and control. This holds true whether the organism engages in behavioral trial and error or cognitively simulates action plans and their anticipated outcomes to direct its actions in insightful ways. In either case, to learn from the process instrumentally, the organism must have a standpoint from which to attend not only to the environmental effects of its action, typically visually, but also to the intended action and the action itself, proprioceptively, and relate them to each other and the goal, and save the relation. This can only happen on an executive tier of functioning that comprehends the intended action, the action, and the environmental effect in a common workspace and representational format that stores the relations. Specifically, it is done by a cognitive monitoring process in which the organism attends to the execution of the intentional action and feeds information back to its store of knowledge about how things are going or have gone.

Another phenomenon that suggests a role for the executive tier of functioning in mammalian learning is curiosity and exploration. Although to my knowledge we do not have systematic data across species, informal observation suggests that mammals are especially curious about things outside of problem-solving contexts, including in play. Just as squirrels store nuts for later consumption, they may store information for later use in problem-solving. One way of thinking about such curiosity is that the organism has learned that learning is useful, which requires an executive tier of functioning in which the organism observes itself learning new things and evaluates that process with respect to the outcomes it generates. Although simpler descriptions of this process are possible, a plausible description is that the organism is learning about the learning process itself, which could only happen, one would think, from an executive tier that comprehends inputs of many different types all in a single workspace and representational format.

Although learning is often opposed to "innate" or hardwired behaviors, the fact is that learning is itself a biological adaptation that may evolve with different features. Boyd and Richerson (1985) have argued and provided evidence that learning as an adaptive strategy arises in a species (or in particular contexts within a species) in the face of environments that change rapidly and unpredictably, such that hardwiring is too rigid to guide effective action consistently. This description accords well with the current account of the ecological conditions in response to which agency in general, and executive functioning in particular, evolved. Thus one hypothesis proposes that highly flexible types of learning—such as the instrumental learning of how actions cause effects—evolved as part and parcel of mammals' skills of executive decision-making and cognitive control in general.

Experiencing One's Own Goal-Directed Action and Attention

With the emergence of mammals some 200 million years ago, then, came a new form of psychological organization: intentional agency as empowered by a new tier of executive functioning. Intentional agents are able to think and plan before they act; that is, in many situations, they are able to cognitively simulate the actions they might perform and to evaluate these action plans to decide which of them to execute and how, such as in some particular sequence, perhaps involving goal embedding. Then, as they proceed to execute the chosen plan, they cognitively monitor the process and keep things on track even if they encounter surprises along the way, for example, by monitoring and adjusting to uncertainty in a decision or by inhibiting any unplanned actions that threaten to disrupt action execution.

This cognitive monitoring also enables them to learn how their actions affect environmental situations, which may be useful either to adjust in the moment or to store for future use. This flexibility and learning evolved so that individuals were able to act effectively in response to whatever novel contingencies might come their way, including those generated by unpredictable instances of social competition.

Researchers in animal cognition have investigated, in one way or another, most of these phenomena: planning, decision-making, inhibition, working memory, and so on. The novelty of my proposal is that I conceptualize these phenomena as different components of a new, integrated tier of functioning, namely, an executive tier operating as its own feedback control system that oversees the operational tier of perception and action, with the goal of facilitating decisions (e.g., in the face of social competition). This executive tier enables the individual organism to direct its actions in new ways (i.e., with plans formulated via cognitive simulations) and to control its actions in new ways as well (i.e., with either-or decision-making and active inhibitory control in which alternative action possibilities are evaluated, and potentially devalued, relative to one another). The executive tier of operation provides for all this new cognitive activity a common workspace and representational format (a.k.a. working memory) that enables the comparative evaluation of simulated actions and results from different perceptual modalities—again, to facilitate better decision-making.

This new form of psychological organization leads, once again, not just to new particular experiences but to a new type of experience. Because reptiles began operating as simple goal-directed agents, they began experiencing the world not just in terms of punctate stimuli but in terms of situations of opportunity and obstacle. Beyond this, operating with an executive psychological tier created for mammals the possibility of experiencing their own perceptual and behavioral functioning. Reptiles and other goal-directed agents do not experience their own perceptions and actions executively, whereas mammals not only experience their own perceptions and actions executively but operate on them from that executive tier. Reptiles and other goal-directed agents are sentient of the outside world; mammals and other intentional agents are conscious of their own actions and perceptions.

Conscious experience thus exists, in my view, only in creatures who operate with an executive tier of functioning, including most mammals and whatever nonmammalian species operate in this way. This proposal is broadly consistent with neuroscientifically based, two-level theories of consciousness, such as that of Graziano (2019), who conceptualizes conscious experience as the organism's cognitive model of its attention to the world, what he calls "the attention schema" (it is thus, minimally, a "higher-order" theory of conscious experience; Brown et al., 2019). But I would replace Graziano's cognitive model with the executive tier itself in its functioning as a feedback control system, whose goal is to facilitate behavioral decision-making on the perception-action (operational) tier. Conceptualizing all of this as emanating from an executive feedback control system also makes clear the many ways in which it goes beyond the simple operations typically attributed to working memory as mainly a capacity limitation. In any case, the key point is that the individual attends from the executive tier to whatever is happening on the operational tier that is relevant to the executive goal of making the best decision, namely, its own goal-directed actions and goal-relevant experiences and how they affect the environment.

Along these lines, Piaget (1976) makes some interesting observations and speculations about consciousness with respect to goal-directed action and experience in general. On the basis of empirical studies with young children, Piaget concludes that the ontogenetically first and most basic objects of conscious experience (as indicated by children's ability to talk about them) lie at what he terms "the periphery" of the process of goal-directed action: namely, on one end, the originating goal (what I was trying to do); and on the other end, the action and its environmental result (what I did, and what happened as a result). Translated to evolution, this might suggest that mammals and other intentional agents, as the simplest conscious beings, experience from the executive tier (i) their own goals, and (ii) their own actions and their effects. This enables them to conceptualize these peripheral components of their goal-directed actions in a single workspace and representational format on the executive tier, which is precisely what enables them to instrumentally learn the relation between goals, actions, and the action's causal effects. In this view, however, it may be that mammals and other intentional agents are not conscious of the more central psychological processes of executive decision-making and cognitive control (i.e., beyond a global feeling of uncertainty in considering a decision); they are doing these things, but they are not conscious that they are doing them. This is an interesting possibility, because, as I speculate further in the next chapter, being conscious of their own executive decision-making and cognitive control—from a second-order executive (reflective) tier—is precisely what great apes, as rational agents, begin to do.

Consciousness is mysterious; so much is clear. But in a fundamental sense, conscious experience is no more—but also no less—mysterious than attention to the external world in the first place. Does anyone really understand how attention—as a means of selectively perceiving some things while ignoring others—actually works? "The nature of consciousness," then, is essentially a question about executive-tier attention to the perception-action tier of psychological functioning, and thus to understand attention is to understand the most basic mechanism generating conscious experience. The human version of conscious experience may have some special qualities, for example, in its use of the perspectives and evaluations of other individuals, its use of language, and its use of sociocultural norms, so we may want to call the human version something different, like self-consciousness. But for now, the essential points are that (i) basic sentience in the sense of attention to, and experience of, the outside world is for agents a psychological primitive; and (ii) basic consciousness involves the organism attending to its own goals, actions, and experience from its executive tier of functioning. My hypothesis is that mammals and other intentional agents are conscious in this sense.

5 Ancient Apes as Rational Agents

The idea of compulsion, as applied to events in nature, is derived from our experience of occasions on which we have compelled [other things]. . . . Causal propositions . . . are descriptions of relations between natural events in anthropomorphic terms.

—R. G. Collingwood, *Essay on Metaphysics*

Great apes are mammals, but humans have always seen them as much more humanlike than other mammals. Queen Victoria referred to the first great ape in the London Zoo (the orangutan Jenny) as "disagreeably human." The Indonesians who first observed orangutans in the forest gave them their name, which translates to "man-of-the-forest." Today scientists in many countries perform invasive research on all kinds of mammals and nonhuman primates, but not on great apes, and the Great Ape Project even advocates for legal rights (and sometimes human rights) for great apes. The explanation for all these special attitudes to great apes is that, compared with other mammals, apes seem so similar to human beings.

I attempt to capture great apes' psychological closeness to humans by referring to them as rational agents. When I call apes "rational," I do not just mean in the economic sense of "pursuing their goals intelligently," which all mammals do, but rather that they operate *logically* and *reflectively*. In terms of logic, great apes do not just experience objects moving and conspecifics acting in space but rather understand in some sense *why* those objects and conspecifics are moving and acting as they are; that is, they understand something of the underlying causal structure of events in the physical world and underlying intentional structure of actions in the social world. Moreover, this causal and intentional understanding is structured by inferences organized into logical paradigms. For example, given causal knowledge that heavy objects destroy termite mounds, if I drop a rock on that termite mound, then it will destroy it. Or given intentional knowl- edge that individuals pursue desired food whose location they know, if my competitor is not pursuing that banana now, then she either does not see it or does not want it. Once an individual has logically connected cause and effect, the individual can now create that effect at will by producing its cause—or imagine doing so—thus opening up a whole new dimension of rational thinking and agentive action.

In terms of reflection, great apes do not just cognitively monitor and control perception and action, as do all mammals; in addition, they cognitively monitor and control executive decision-making itself. They do so by using a second-order executive tier of functioning, what I call the reflective tier, in which individuals monitor and evaluate their own first-order executive decision-making and cognitive control. This new executive tier enables apes not only to identify but also to intervene and correct problems in their first-order executive functioning. The reflective tier is also responsible for great apes' ability to attribute mental states to others, as it provides the individual with access to its own mental states (metacognition), as well as a second-order workspace and representational format in which to make the relevant comparisons. With logically structured causal and intentional inferences and second-order executive (reflective) control, we now have what may legitimately be called *rational* thinking and decision-making (even if it is not yet normatively rational in a humanlike way).

In support of these special cognitive skills, great apes have evolved extremely large brains (however this is measured). Primate brains in general are comparatively large among mammals, subdivided into more functionally distinct areas, and characterized by neurons that are both larger in size and packed more densely than in other mammals. Great ape brains, in particular, are even larger than those of other primates, and the corpus callosum (operating between the hemispheres) has thicker axons capable of faster transmission (Kaas, 2013). Compared with other mammals and primates, great apes also have an expanded prefrontal cortex, the well-known seat of all kinds of executive functioning (Smaers et al., 2017).

I am making an extremely large leap here from mammals to great apes, and indeed, great apes almost certainly represent the end point of a protracted evolutionary process, with various species of ancestral primates in

Figure 5.1
Imagined early great ape twenty million years ago.

between. But with specific reference to the rational processes on which I am focusing here, we do not have sufficient experimental data with monkeys to identify precisely the ways in which they resemble or differ from apes. So even though the evolutionary leap to great apes almost certainly had smaller steps along the way, which might be represented in extant monkey species, for current purposes, I will focus on humans' nearest great ape cousins. Within great apes, the species about whom we know by far the most is chimpanzees, who just happen to be, along with bonobos, humans' closest living relatives. Although the great ape species differ from one another in significant ways, their agentive behavioral organization would seem to be broadly similar, so I will proceed with chimpanzees as representative of great apes in general.

Socioecological Challenges

The first great apes emerged about twenty million years ago. Soon dozens of ape species were roaming throughout Africa and Eurasia. By most accounts, the last common ancestor to the extant great apes was somewhat different from other primate species, such as monkeys.

Early Miocene apes left Africa because of a new adaptation in their jaws and teeth that allowed them to exploit a diversity of ecological settings. Eurasian great apes evolved an array of skeletal adaptations that permitted them to live in varied environments as well as large brains to grapple with complex social and ecological challenges. These modifications made it possible for a few of them to survive the dramatic climate changes that took place at the end of the Miocene. (Begun, 2003, p. 80)

The five extant great ape species are orangutans, gorillas, bonobos, chimpanzees, and humans. (From this point on, I use the terms "great apes" and "apes" to mean nonhuman great apes.) Extant great apes have retained their ancestral psychological adaptations for diverse and flexible foraging environments. However, at some point before the extant species diverged, there arose a new set of socioecological challenges, prompted by apes' growing predilection for fruits (all contemporary great apes have a strong preference for fruits). Fruits mostly grow in widely dispersed patches or clumps, that is, in trees, which have only limited avenues of access. Clumped resources with limited avenues of access create situations of especially intense social competition among all who approach them at the same time, with the possibility that some individuals may try to monopolize them. This new socioecological challenge led to major changes in the ways that great apes' social groups were structured. Individual apes no longer foraged in large groups, as did other cercopithecine primates, but rather split into smaller foraging parties. Contemporary chimpanzees and bonobos still live in largish social groups for purposes of sleeping and group defense, but they forage daily in small bands that fission and fuse in many and various ways throughout the day. My hypothesis is that such fission-fusion societies represent the original social organization of the first great apes. Contemporary orangutans and gorillas are the exceptions that prove the rule, because they live in partial solitude or in single-male groups (the fissions became semipermanent), which also reduces the size of the foraging party and thus the social competition.

In addition to a new type of social organization, great apes' especially intense food competition led them to a new type of psychological organization (see Sterelny, 2004, for this argument for humans, which applies to apes in general). For one thing, their cognitive simulations and thinking became logically organized. In the physical domain, great apes are well known to be uniquely skilled among primates and other mammals at making and using tools, underlain by special skills of causal understanding (generalized even to causal forces external to the self). In the social domain, great apes are well known to be uniquely skilled among primates and other mammals in social learning and intentional gestural communication, underlain by special skills of intentional understanding (or "mind reading"). Understanding causality and intentionality are key cognitive skills for rational agents because they expand the field of agentive action to include not just events and actions in the world but the underlying causes of those events and actions (and their logical interrelations), which can potentially be manipulated to produce desired effects. In addition to these new forms of logical thinking, great apes' especially intense competition for food led them to reflective forms of decision-making and cognitive control; that is, they began to make more efficient decisions by employing a second-order tier of executive (reflective) functioning, creating new ways of planning, choosing, and controlling their actions. Operating logically and reflectively in these ways means that ancient great apes evolved a new organization for psychological functioning: rational agency.

Understanding Causal Events

One way that great apes can deal with their especially intense competition for food is by using tools to extract nutrients that others are ignoring. Great apes' use of tools is not totally unprecedented among mammals and other primates, but they use their tools in unprecedented ways in activities such as fishing for termites and cracking nuts. Although only chimpanzees and orangutans need tools to forage in the wild, all great apes can, in the right circumstances, use a wide variety of tools in a highly flexible manner, including learning to use novel tools quickly and proficiently in experiments.

Organisms that understand causality understand not just what is happening but also, to some extent, why it is happening, which creates the possibility of manipulating the cause to produce the effect. Thus, in experiments in which great apes must *choose* a tool that is causally appropriate for a novel problem (e.g., one that is long and rigid, not short or soft), they seem to understand the causality involved, as they make the right choice even though they have never before encountered these tools, and the novel problem is completely out of sight. To do so, they must be assessing the causal relation of the physical characteristics of the tool (as an enabling cause) to the physical characteristics of the problem substrate (as they are imagining it; Manrique et al., 2010). Great apes' causal understanding of tools leads to considerable behavioral flexibility in using them.

In addition, when faced with a novel physical problem, great apes can also take control of the causal process and *make* new tools that will work in the new context. In this case, they are first imagining an effect that is needed to solve the problem, and then going back to create a cause. For example, in the wild, chimpanzees routinely modify too-leafy branches by stripping leaves from them so that they will fit into termite holes (McGrew, 2010). In captivity, some great apes can even turn water into a tool. Thus, when a peanut is at the bottom of a narrow tube, apes will go to a water source, get water in their mouths, and spit it into the tube multiple times to get the peanut to float to the top so that they can reach it (Mendes et al., 2007). Another creative instance involves orangutan mothers physically pushing their infant up to an opening that is too narrow for adult hands to penetrate so that the infant will reach through and get a piece of food (which the mother then immediately appropriates; Völter et al., 2015). In engaging in toolmaking of this kind, great apes are using their understanding of the causal relations involved to actually intervene in the process agentively.

But beyond exploiting the causal relations involved in using and making tools—that is, exploiting tool properties as enabling causes—great apes can, in some contexts, understand causal forces that operate totally independently of their own actions. For example, in one study, apes attempted to identify which of several opaque bottles contained juice (and they could choose only one). The apes quickly alighted on the strategy of picking up the bottles one by one to test for their weight, and as soon as they found a heavy one, they chose it. (In contrast, if the bottles all weighed the same, but the bottle with juice was painted red, the apes found it extremely difficult to associate the red color with the presence of juice across trials.) They thus understood, in some sense, that the bottle's extra weight was caused by the juice (Hanus & Call, 2008). In a study in which they could not act on the objects at all, chimpanzees inferred that when one end of a balance beam tilted down, it meant that the opaque cup on its end contained a banana (which they did not infer if someone pushed and held that end of the balance beam down), indicating an understanding that heavy things exert a downward causal force (Hanus & Call, 2011).

Again, when apes are assessing causal relations totally external to them-selves, they understand not just what is happening but also why it is hap-pening, so they have the possibility of manipulating the cause to produce the effect. Thus, in one experiment, chimpanzees observed a human press-ing a button in two different conditions: either she pressed the "causal button," which was followed by the immediate delivery of juice (cause-then-effect), or she pressed the "noncausal button" only after the delivery of juice (effect-then-cause). When given the opportunity to produce the juice themselves, the chimpanzees pressed the causal button already on the first trial (Tennie et al., 2019). Similarly, Völter et al. (2016) had great apes observe humans placing various objects on a "blicket machine" that sometimes led to food being dispensed. Some actions were causally effec-tive, and some were not. When given an opportunity for themselves, the apes made good use of what they had observed, and intervened in the pro-cess in appropriate ways to make the machine dispense the food. Finally, in some rare instances, apes even attempt to intervene to discover causes. Thus, when chimpanzees were rewarded for taking overturned blocks and setting them upright, and then one of the blocks would not stand upright (it was asymmetrically weighted inside), some of the apes picked up the block and visually inspected it underneath, seemingly trying to discover the cause of the problem (Povinelli & Dunphy-Lelii, 2001). Great apes thus learn not only from actions and their results, as do most mammals, but also from the causal and intentional relations *among external entities*. In great apes, instrumental learning means causal learning.

Great apes' causal understanding generates creative inferences organized into logical paradigms. For instance, in the experiments on tool choice, apes infer things such as "if a tool with property A is used, then B must happen." Then, actually using the tool completes the following syllogism: (i) if A is used, then B happens; (ii) A is used; (iii) therefore B should hap-pen. In other experiments, apes can make backward-facing inferences from effect to cause, in this case using exclusion based on a simple kind of nega-tion (what logicians call contraries). Thus Call (2004) had an experimenter show a chimpanzee a piece of food, which was then hidden in one of two cups. Next, in the key condition, the experimenter shook the empty cup. The chimpanzee observed only silence. To locate the food, apes had to infer backward in the causal chain to why that might be, namely, that no food was inside the cup. The chain of inferences was thus something like the following: (i) the shaking cup is silent; (ii) if the food were inside the shaking cup, then it would make noise; (iii) therefore the food must not be in this cup (but rather in the other one). Following José Bermudez's (2003) analysis, these inferences and paradigms involve the two key elements of logical thinking: the if-then conditional and negation. Both occur in only "proto" form: the if-then conditional is proto because it only concerns causal (not formal) relations, and the negation is proto because it only concerns contraries such as presence–absence, noise–silence, and so on. This proto-logic can still be realized in image-schematic representations involving causal forces and mutually exclusive (contrary) situations.

Great apes thus seem to understand the underlying causal structure of their physical worlds in ways that other mammals do not. (Although other mammalian and primate species have not been tested in all the relevant experimental paradigms, the fact that none of them uses or makes tools in the wild in anything like the great ape way suggests that great apes do indeed have special cognitive abilities of this type.) And they see these causal relations as logically interrelated—the physical world can be rationally understood—enabling them to create effects by manipulating causes.

Understanding Intentional Actions

Another way that great apes can deal with their especially intense competition for food is by outsmarting their competitors, who, as intentional agents, initiate their own actions, often unpredictably and in ways aimed at making life difficult for others. So our question now is whether great apes understand what "causes" the actions of others; that is, whether they understand that other agents operate as feedback control systems, whose actions are directed toward their goals as guided by their perceptions.

In numerous experimental situations, great apes predict the actions of others based on an understanding of their specific goals and perceptions in that context. In terms of goals, chimpanzees in experiments discriminate reliably between actions done intentionally toward a goal or accidentally. For instance, when a human is sitting in front of a chimpanzee holding food and never gives it to her, the chimpanzee behaves differently depending on whether the human seems to be intending to give her the food or not (e.g., is in the process of trying and failing; Call et al., 2004). In terms of perception, when a subordinate chimpanzee is competing for food with a dominant, it can take into account whether or not that dominant sees a potentially contested piece of food (because of judiciously placed barriers). And the subordinate chimpanzee can even tell if the dominant has seen the food in that location in the immediate past and thus knows it is there (even though at the moment the dominant cannot see it; Hare et al., 2000, 2001). In general, in such experiments, apes understand that a competitor will go for an object only if (i) the competitor wants or has a goal or desire for that object (i.e., it would not compete for a rock); and (ii) it perceives or knows that the object is in a certain location. Great apes thus understand how competitors work as agents—that is, in terms of their goal and perception—and can use this understanding in novel contexts to predict their behavior.

In addition, at least some great apes show intentional understanding in their social learning. Much of apes' social learning is emulation learning, in which the learner notices an effect in the environment that has been produced by a conspecific, and then reproduces that effect using her own means. But after a few month of training, young chimpanzees can learn to match their own actions to a human's novel actions, that is, to imitate (Custance et al., 1995). But beyond reproducing outcomes and actions, Tomasello and Carpenter (2005) found that three human-raised chimpanzees selectively reproduced actions that a human demonstrator intended to perform over actions she performed only accidentally; the chimpanzees also performed actions that a human intended to perform but did not actually succeed in performing. The apes thus aligned their intentions with those of another agent. Even more impressive, Buttelmann et al. (2007) found that human-raised chimpanzees did not imitate a human demonstrator performing a strange action, such as turning on a light with his foot when he had no other choice, since his hands were otherwise occupied; they did not because his decision-making situation was different from theirs (they had no constraints). But the chimpanzees did imitate him when he had freely chosen the same strange action in the absence of constraints (he and they were similarly unconstrained). This process is called "rational imitation" because the social learner is comparing its own process of decision-making to that of another agent (see also Buttelmann et al., 2008).

Beyond just predicting what others will do and socially learning their intentional actions, in competitive contexts, apes sometimes attempt to actively manipulate what others perceive so as to affect what they do. Thus, in one experiment, chimpanzees chose to approach a contested piece of food from a route that concealed their approach from a competitor (Melis, Call, & Tomasello, 2006; see also Karg et al., 2015a). This strategy is analogous to the way that chimpanzees manipulate physical causes: they understand what causes an effect—in this case, they understand that if a competitor sees their approach, it will take the food first—so they take control of the cause, in this case by manipulating (concealing) what the other can see. Importantly, in another experiment in the Melis et al. study, apes approached contested food via a quiet (versus loud) route so that their competitor did not hear them, demonstrating flexible and general skills in manipulating others' perceptions.

In a similar manner, great apes learn and flexibly use communicative gestures to manipulate the goals and perceptions of others in less competitive contexts such as play, sex, grooming, and group travel. Apes learn such gestures and produce them intentionally in a way that is unique among mammals. One example is when a youngster wants another to play. Initially the two apes begin by playfully hitting each other. From these interactions, an individual may ritualize an "arm-raise" gesture: it raises its arm as a kind of playful threat to hit, which the other understands to be a play initiation. This gesture and many others like it are learned through a process that once again suggests taking control of a causal link. Analogous to making tools and concealing one's location, the apes' strategy is formulated from effect to cause: I want her to play (effect), and raising my arm will make her *see* me and *want* to play (cause), so that is what I will do (see Halina et al., 2013, for a detailed study of this process). Experimental studies also demonstrate that apes know that the recipient must perceive the gesture for it to work. Thus, when apes produce a visually based gesture, like an arm raise, they only do so when the other is looking (which they do not do for physically based gestures; Liebal et al., 2004), and they produce various "attention getters" whose only function is to get the recipient to attend to them and what they are doing (Call & Tomasello, 2007). Apes thus understand key aspects of how their agentively produced action, the gesture as tool, affects others psychologically by causing changes in their goals and perceptions (see also Bohn et al., 2016).

As in the physical/causal domain, apes' inferences in the social/intentional domain are logically structured. Consider the food competition experiment cited earlier. The competitors in this situation inferred the following of one another: if he has the goal of getting the food, and he perceives its location (and so knows how to get it), then he will go for it. But if he does not have the goal, or he does not perceive a way to achieve it, then he will not pursue the goal. And from the other direction: if my competitor is engaged in a particular activity, then she must have had both a relevant goal and a relevant perception. Again, these inferences constitute a kind of logically structured paradigm analogous to the human practical syllogism. Perhaps most impressively, in the Buttelmann et al. (2007) rational imitation study, apes made a backward-facing exclusion inference based on proto-negation. Specifically, when they saw a human operate a device with his foot when his hands were externally constrained, the apes inferred from his behavior backward to his decision-making: (i) he is not using his hands; (ii) normally, if he had a free choice, he would be using his hands; (iii) therefore he must not have a free choice (so I can ignore his action choice). As in the case of logically structured causal inferences in the physical domain, then, these logically structured inferences about others' intentions and actions employ the two most basic elements of human logical thinking: conditional (if-then) inferences and proto-negation (based on contraries).

Great apes thus also seem to understand the underlying intentional structure of their social worlds in ways that other mammals do not. (Again, although other mammalian and primate species have not been tested in many of these paradigms, the fact that none of them socially learns or communicates intentionally in the wild in anything like the great ape way suggests that great apes do indeed have special cognitive abilities of this type.) And they see these intentional relations as logically interrelated—the social world can be rationally understood—enabling them to create effects by manipulating causes.

Rational Decision-Making and Cognitive Control

Great apes' understanding of the world is thus logically structured in both the physical and the social domains. The individual ape not only can imagine what would happen if it performed a certain action, as do squirrels, but also can make logical inferences and therefore predictions about what would happen if things acted on one another, or even did not act on one another, in the external world. This pertains not only to inanimate objects bumping into one another in causal ways, but also to agents acting on things or other creatures intentionally. In addition, as we shall now see, great apes' decision-making and cognitive control are reflectively organized. That is to say, great apes have come to deal with their especially intense competition for food not only by coming to understand how the world works but also by reflectively planning, choosing, and controlling their own actions

Like all mammals, great apes plan their actions. For example, like rats, apes in experiments can travel from one differently rewarded foraging location to another across multiple locations in an efficient manner. But in such situations, apes could be using various heuristics that make the planning involved minimal. Völter and Call (2014), therefore, gave apes a more complex task in which they had to mentally plan a sequence of actions before executing the first action. Specifically, apes were presented with a vertical maze where dropping a ball in a particular hole (of several) at the top would commit them to a sequence of routes farther down in the maze, some of which were dead ends. So the apes had to mentally trace out the causal structure of various routes, several levels deep, to find a successful route to the bottom before choosing the beginning hole. They were making what could be called a rational plan because it was based on logically structured causal inferences about what might happen to the ball when it encountered certain passages and obstacles in the maze. The apes learned to imagine such rational plans as quickly and accurately as five-year-old human children.

But this is planning a means to a goal that is already activated. In a spectacular experiment, Mulcahy and Call (2006) gave bonobos and orangutans the opportunity to plan for a future goal that they did not at the moment actually have. Once per day, apes learned to use a tool (of several possible tools) to retrieve a single reward from an apparatus in a room. Then they were sent out of the room. When they returned, the apparatus was rebaited, but there was no tool to be found. The apes quickly learned within a few trials to select the appropriate tool and take it with them as they were ushered out of the room, so that they could have it later when they reentered. Thus they anticipated that they would need the tool later, took it with them and saved it for up to fourteen hours outside, and then brought it back into the test room and used it successfully when they later had access. Anticipating having a goal in the future requires a kind of reflective understanding of goals and their role in the planning process. Planning for a future imagined goal in this way would seem to require some new executive, that is reflective, cognitive skills.

In terms of the evaluation and choice of action plans, almost all the research with apes has focused on risk profiles that characterize the way that different species weigh reward value in comparison with reward probability (e.g., Heilbronner et al., 2008). In addition, two recent studies have shown that great apes can distinguish more generally between situations of risk (when they can judge the likelihood that they will make a correct decision) and uncertainty or ambiguity (when they have no relevant information about this likelihood; Rosati & Hare, 2011; Romain et al., 2021). But most important in the current context is not species-wide tendencies but rather individual agency. The point is that great ape individuals make decisions differently depending both on context and on their own internal state. For example, Haun et al. (2011) found that all four great ape species modulated their decision-making within a single experiment depending on risk context, tending to choose the safer option more often as the risks of failure increased over trials. In terms of internal state, chimpanzees in the wild engage in the high-risk activity of monkey hunting more often in times when many plant and fruit resources are also available, presumably so that in case of hunting failure, the chimpanzees have adequate backup options (Gilby & Wrangham, 2007). Individual great apes thus make decisions that are both context sensitive and state sensitive, thus exercising their individual agency (for reviews, see Rosati & Stevens, 2009; Rosati, 2017c).

But perhaps the greatest difference in mammals' and great apes' decision-making comes in the way that individuals monitor and control their uncertainty. Recall that individual rats can decide to avoid an uncertain option with a potentially big reward and opt for a safer option with a lower reward. Apes do this too, but they also do something more: they not only executively monitor their uncertainty but also reflectively monitor and control the decision-making process itself. They do this by seeking to identify the cause of the uncertainty and doing something about it if they can. Apes have unique skills of what is sometimes called metacognition, using these unique skills to effect a better, perhaps rational, process of decision-making.

Call and Carpenter (2001; see also Call, 2010) created a variation on the opt-out uncertainty paradigm for great apes. The difference in this paradigm is that merely feeling uncertain and going for the only other option is not enough; here the individual must diagnose why it is feeling uncertain and do something about it. The method went as follows: Chimpanzees either did or did not see the process of food being hidden inside one of several tubes. When they witnessed the hiding process, they chose a tube immediately. But when they did not witness the hiding process, they went to some trouble to look into the tubes to discover where the food was located before choosing. The apes knew when they did not know, or at least when they were uncertain; but in this case, they reflected on the decision-making process to identify the problem and fashion a solution (for similar studies and results with rhesus monkeys, see Basile et al., 2015; Rosati & Santos, 2016; for negative findings with nonprimate mammals, see Roberts et al., 2012). The apes diagnosed that they were missing some information and then determined how to alleviate their ignorance. This behavior is not just a trivial function of animals being used to looking for food before obtaining it, as even otherwise very clever species, such as capuchin monkeys, always look for the food before they choose, even on trials when they have seen the hiding process. In a variation on this theme, Bohn et al. (2017) found that when apes needed a tool of a certain type and did not know its location, they would seek that information before acting. Seeking information to facilitate better decision-making means not only that great apes are reflectively monitoring and controlling the decision-making process, but also that they are employing a kind of computational rationality in the sense that they must decide if the potentially available information is worth the effort needed to gather it.

One further study justifies even more strongly the designation of great ape cognitive self-monitoring and control as rational. O'Madagain et al. (submitted) gave apes the opportunity to visually locate the best food in a situation at location X. The apes did this, indicating their belief by choosing that location (though not receiving the food as a result). Then they were exposed to new information that called their initial belief into question: the new information suggested that the best food might be in location Y. The apes had the possibility at this point to seek further information (or not) that could either confirm or disconfirm their initial belief. Many apes then actively sought more information to resolve the discrepancy between their original belief and the new information, by looking again into location X to check their initial judgment a second time, so as to make the best decision. The apes were in this case self-monitoring and controlling their executive decision-making after they had already made an initial decision;

they were reflecting on the decision in the light of newly obtained information and discerning the need to possibly revise that decision. Attempting to causally diagnose problematic decisions before they are behaviorally executed fulfills a standard criterion for rationality: self-critical reflection on one's own thinking and decision-making.

Finally, at the point of action execution, great apes also display especially strong skills of inhibitory control as compared with other mammals and primates. In two studies reported in the previous chapter, E. MacLean et al. (2014) found that all four species of great ape outperformed rodents, carnivores, and elephants in two tasks of inhibition, and Amici et al. (2008) found that three of the four great ape species were especially good in five different tasks of inhibition as compared to other primates. But of special importance for issues of reflective cognitive control are situations in which a conflict exists between one goal and another (each determining attention to a different set of relevant situations). Thus Herrmann, Misch, Hernandez-Lloreda, and Tomasello (2015) gave a battery of tests involving such goal and attentional conflicts to chimpanzees as well as human children of two different ages. Surprisingly, the apes turned out to be as skillful as the three-year-old children (though less so than the six-year-old children). The apes were able to

- inhibit taking a closer, smaller reward to pursue a larger, more distant reward (spatial discounting; see Rosati et al., 2007, for a study of ape versus human temporal discounting);
- inhibit a previously successful action in favor of a new one demanded by a changed situation (strategy updating);
- persist to a goal through failures and in the face of temptations to do something different (behavioral persistence); and
- concentrate through distractions in the face of temptations to attend to something else (attentional focus).

The latter two tasks involve the ability to maintain goal pursuit and its associated attention, even though other attractive goals or stimuli are clamoring, bottom up, for actions or attention. Chimpanzees have shown further evidence of the ability to deal with conflicting attentional demands in a study using a modified Stroop task (Allritz et al., 2015). Having learned to touch a square frame of a certain color for a reward, if an image inside that frame was now a different color, the chimpanzees could still be successful (even though it now took them longer to make a choice, and they made more mistakes than if the image and frame were the same color). But the most impressive skill chimpanzees displayed in the Herrmann et al. (2015) study was to deal with a situation involving conflicting goals. In a final task, the chimpanzees were able to make themselves do something fearful for a highly desirable reward at the end, requiring them to somehow harmonize the goals of getting the reward and not doing anything too dangerous. The apes mostly were successful at adjudicating between their conflicting goals—they chose to pursue the desired reward and to ignore the danger— but sometimes only after significant hesitation, indicating that they did indeed experience the goal conflict.

Perhaps chimpanzees' most impressive feats in the cognitive control of attention and action may be seen in two continuous performance tasks (CPTs). Herrmann and Tomasello (2015) gave chimpanzees and human children at four to five years of age two different CPT tasks in which they had to simultaneously monitor two constantly changing locations where rewards might appear. In one task, they had to simultaneously monitor two apparatuses dispensing rewards asynchronously but at roughly the same rate; in a second task, the primary apparatus was dispensing rewards continuously at a fast pace, but then bigger and better rewards appeared intermittently and unpredictably at a second location. Surprisingly, chimpanzees were skillful in both tasks, maximizing their rewards as skillfully as the children. Continuous monitoring tasks require individuals to maintain two goals and their associated attentional demands simultaneously and switch between them flexibly as the situation demands so as to maximize rewards. This kind of proactive cognitive control demonstrates an ability to monitor the decision-making process for possible incompatibilities between goals before acting, another criterion that is often used to designate decision-making as rational. And, again, their skills in CPTs might suggest that apes employ a kind of computational rationality, in this case to decide how to negotiate between their conflicting goals and deploy their attention in the most efficient manner.

The Reflective Tier and Its Experiential Niche

The ability to monitor, troubleshoot, and intervene in processes of executive decision-making and cognitive control would seem to require another (reflective) tier of executive functioning on top of the normal mammalian executive tier. Squirrels and other mammals are executively self-regulating their intentional actions. But whereas they operate *with* processes of executive decision-making, in the absence of a second-order executive tier, they cannot operate *on* them, so they cannot monitor and control them. In contrast, the studies I have cited suggest that great apes can monitor and control not only their goal-directed actions but also the processes involved in their own executive decision-making and cognitive control (and possibly make decisions about efficiency or so-called computational rationality). I attempt to capture the overall organizational architecture involved in such rational (reflective) self-monitoring and decision-making, in a very general way, in figure 5.2 (which is just figure 4.2 with a reflective tier on top and attention to causal and intentional relations in the environment added on).

I further hypothesize that this reflective tier of agentive organization was also instrumental in the evolution of great apes' unique cognitive skills for understanding causal and intentional relations in the external world. Specifically, I propose that great apes' understanding of causality and intentionality resulted from an attribution to external events of some of their own decision-making processes that they were now able to access from their new reflective tier of functioning. And the new tier of reflective functioning also provided the common workspace and representational format necessary for comparing and aligning internal (first-person) and external (third-person) events in the attribution process. In other words, the emergence of a reflective tier of executive functioning enabled great apes to attribute causal and intentional relations to outside entities and events because (i) it enabled the individual to attend to its own executive decision-making, and (ii) it created a common workspace and representational format for aligning self with outside other so as to make the attribution. The way this worked was similar but slightly different for intentionality and causality.

Beginning with the "easier" case, great apes understand other agents as intentional agents acting and making decisions toward goals as guided by perceptions. The proposal that this understanding originates evolutionarily with self-experience is a variant of so-called simulation theory (Gordon, in press). The point is a conceptual one. If a Martian came down to earth and informed me that without any obvious organs it could still "see" things, how could I understand what it means to "see" something except through my own experience of seeing? Similarly, if the Martian said it "desired" certain "goals," how could I understand what it means to desire a goal except through my own experience of desiring goals? But beyond simulation, an

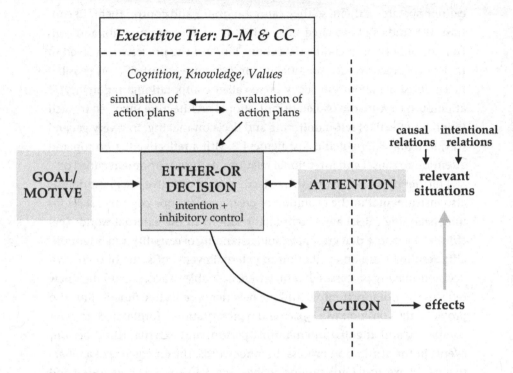

Figure 5.2
Great apes' rational decision-making via two tiers of executive control. Shaded components are analogous to, or the same as, those for intentional agents. Dark-shaded component at the top represents apes' unique second-order (reflective) tier of executive decision-making (D-M) and self-regulation (S-R).

element of "theory" is involved as well. That is, the other agent's particular goals and experience are different from mine in specific circumstances, and to understand them on any given occasion, I must have hypotheses or theories about its particular goals and perception on this occasion: perhaps it has access to food that I do not, or perhaps it is seeing something that I am not. One account of this general type is what has been called in the computational modeling literature "inverse planning" (e.g., Baker et al., 2009): whereas in my own action planning, I start with a goal, figure out the means, and observe the result, in understanding others' intentional actions, I see the result and the behaviors used as means and must infer the goal and supporting perceptions. Thus great apes make predictions about others' actions both by attributing self-based concepts to them and by forming hypotheses about the others' particular mental states given the particular situation they face at the moment.

Experimental studies support the proposal that great apes attribute their own experience to others. Karg et al. (2015b) arranged for a chimpanzee subject to experience a situation in which it could see through a screen lid on a box to detect what was inside. The box was then reoriented, so that now, from the ape's new side-viewing angle, the screen lid was opaque. A competitor then approached the box and looked straight into it, from the angle that the chimpanzee subject had used originally. When the two of them now competed for the food inside the box, the chimpanzee subject knew that the competitor could see through the lid to the food inside, even though she herself could not see the food at the moment. The only way the subject could know this was from her own previous experience of having looked directly through the lid into the box from the original angle, which she was now attributing to the competitor. Kano et al. (2019) report similar findings using other ape subjects and a different (anticipatory looking) methodology. And in another experimental design completely, Schmelz et al. (2013) had two chimpanzees compete with each other for food. To be successful, an individual had to predict which food a conspecific would prefer and so choose. Both chimpanzees predicted the other's choice by attributing to him a preference for the food that they themselves preferred; that is, they attributed their own specific psychological state of preference to their competitor.

Attributing my own manner of functioning to another creature also requires that I see enough similarity between the two of us to justify the attribution. Evolutionarily, it is possible that the ability to attribute one's own psychological functioning to others evolved in the context of great apes' especially intense food competition: attributing my manner of functioning to a competitor facilitates my ability to predict her behavior (and so to win the competition). But more likely, in my view, the adaptive contexts that were most directly responsible were those involving social learning and imitation, because in these situations, individuals need to simulate others' psychological functioning as compared with their own, so as to align their actions and intentions with them. This hypothesis is supported by the fact that many mammals engage in fairly intense food competition, but this has not led them, apparently, to understand others as intentional agents in the same way as apes. So perhaps apes' unique skills of social learning and imitation, requiring alignment of self and other at one or another psychological level, evolved after the split with other mammals and so provided a unique foundation for apes to engage in all other types of self-other alignment. A coevolutionary process was almost certainly at work here, as the need for better simulations of others in competition, and the need for better simulations of others in social learning, facilitated and benefited from each other.

The process of attributing mental states to other agents based on one's own mental states is almost certainly a skill unique to primates among mammals (indeed, some monkey species show some skills; e.g., Santos et al., 2006). But it is at least relatively straightforward because there is a clear similarity between the actions of self and other: all individuals of the same species, including the self, are highly similar in their bodies and behavior. But the generalization from self to other is not nearly so straightforward when considering attributions to physical events involving inanimate objects and physical causality. Unlike animate agents, physical objects only move when they are "forced" or caused to move by an animate agent—or else by some mysterious action at a distance like gravity (which Isaac Newton himself considered an occult force). David Hume thus wondered about the basis of human causal understanding. When one billiard ball strikes another and knocks it across the table, we experience only a spatiotemporal contiguity: a moving ball contacts a stationary ball, and it then moves, seemingly as a result. But what justifies an inference that a causal "force" is involved?

Recall that Dickinson (2001) presented evidence that rats do not just associate their act with its result but also understand that their act caused the result. However, a huge gap separates such internally generated causality and external causality among inanimate objects. How can this gap be bridged? Piaget's (1952) idea is that the bridge is none other than the use of tools, which, as we have seen, is a skill at which apes excel. To use a tool flexibly and reliably there must be an integration of the movement of the tool, as caused by the agent, and the properties of the tool in relation to the substrate. Therefore the cause of successful tool use is *both* the organism's action and the properties of the tool in relation to the substrate, across the organism-environment divide, as it were. But tool use just concerns the causal properties of the tool as passively participating in the process; the tool properties are what we have called enabling causes. Understanding objects as exerting a causal force independent of the self's own actions requires a still further step. For this, it may be that the ape needs to somehow see physical objects as operating in the manner of intentional agents, that is, in analogy with the causal relations that hold between an agent's action and its effect in the environment. Perhaps apes are making some such animistic attribution to physical events, and this is the basis for their understanding of physical causality (see Collingwood's epigraph to this chapter).

Interesting evidence for this proposal comes from the fact that great apes structure their causal understanding into paradigms of logical inferences, as I have described. If they know that event X causes event Y, then they know that if X happened, then Y did also, and also that if Y did not happen, then X did not either. Such logically structured inferential paradigms constitute evidence for a self-based hypothesis for the origins of causal attribution because they almost certainly derive from the causal logic of the agent's own action. Thus the kind of causal understanding of one's own action that Dickinson attributes to rats yields inferences such as the following: if I act, there will be a result; if I do not act, there will not be a result; if there is no result, then I did not act causally effectively; if there are only two ways to cause a result, and the first one is not causally effective, then the other one will be causally effective; and so on. These kinds of inferences are made on the first executive level about one's own actions and their effects on the operational level (as already in rats). Then, from the reflective tier, great apes are able to align these internal causal inferences about self-action with external events that seem self-generated (e.g., objects spontaneously fall or are blocked), building on their existing understanding of tool properties as part of the causal sequence. Channeling Piaget (1974), then, we may say that an ape's inferences about the causes of its own actions are implications, whereas its attempts to explain external events (e.g., so as to predict them) are explications, both requiring a reflective understanding. They both use the same "logic of action," just differently.

Obviously, attributing the causality of one's own actions to the behavior of inanimate entities in the external world is a much less direct process than doing so in the case of other intentional agents. Nevertheless, it is noteworthy that young human children (as well as adults in many societies) have a strong tendency to explain physical events animistically (e.g., clouds move themselves as agents, or winds are agentive causes) or anthropomorphically (e.g., imaginary or humanlike beings or deities make things happen). And our own intuitions about how gravity "pulls" things down or how billiard balls "push" one another around are most plausibly based on analogies to the forces that we ourselves create and use in pulling and pushing things—what else?

These hypotheses about the conceptual bases for great apes' understanding of causal and intentional relations in the external world are speculative, but I see no alternative if we want to tie these cognitive skills to apes' unique forms of reflective (rational) decision-making and cognitive control. Lizards and other reptiles know nothing of causality or intentionality because they are only operating with a primary tier of attention and action. Squirrels and other mammals understand the causality of their own actions on the world because they operate with an executive tier of functioning. And apes understand the causality and intentionality of events in the world because they operate with a second-order reflective tier of executive functioning that gives them metacognitive access to their own decision-making and cognitive control, which they then can attribute to the external world.

These hypotheses thus represent one more example of my claim that changes in agentive organization lead to changes in the organism's experiential niche. In this case, the change in agentive organization characteristic of great apes—the emergence of a second-order tier of executive decision-making and control—led to the formation of an experiential niche structured by the causes underlying physical events and the intentions underlying agentive action, both organized into similar logical-inferential paradigms, enabling individuals to imagine causally and intentionally structured states of the world that are not directly perceived.

But Are They Really Rational?

And so my claim is that great apes are rational agents. They plan for future goals that they are not now experiencing; they make logically organized inferences based on an understanding of external causal and intentional relations explaining why things happen; they are self-critical of their own decision-making processes, causally analyzing problems or conflicts and intervening to resolve them; and they display impressive skills of inhibitory control and the resolution of goal conflicts at the stage of action execution. I propose that they could not do all these things with the psychological organization of mammals in general; rather, apes need more and better executive processes. If mammals' first-order executive tier of agentive functioning aims at better decisions, great apes' second-order executive tier of agentive functioning aims at *even better* better decisions.

One may, of course, argue about whether all this sophisticated functioning justifies the appellation "rational." As is well known, the term "rational" is used in many diverse ways in the social and cognitive sciences and philosophy. Great apes are clearly rational agents in the sense of the economists who only require them to pursue their goals and preferences intelligently (see the study of Jensen et al., 2007, demonstrating that great apes are rational maximizers in an ultimatum game). Further, although we do not have all the necessary data, it is likely that great apes also employ some kind of "computational rationality," as defined by computational modelers, to be efficient in their decision-making by taking into account the various costs involved. However, great apes are not rational agents in the strict philosophical sense that they regulate their thinking by socially normative standards of rationality; this is almost certainly a uniquely human capacity, as we shall see in the next chapter. The criteria for rationality that I have used here aim at a middle ground by employing criteria often used by philosophers in addition to the use of socially normative standards. In particular, I have used as criteria (i) thinking about the external world using logically structured causal and intentional inferences, providing rational coherence to experience; and (ii) adopting a reflective and self-critical stance to one's own thinking and decision-making, including adjudicating between conflicting goals before acting by reflecting on their relative merits, providing rational coherence to one's psychological functioning in general.

My specific model of great ape rational agency borrows components from various research traditions in the cognitive sciences. The basic structure of the model for all agentive species is the kind of feedback control system common to many different theoretical approaches to intelligent action. Decision-making in the model comprises a kind of generic version of processes common to many different approaches in decision science, including those employing simplifying heuristics. And the processes of executive or cognitive control in the model are again generic versions of the models used in various branches of cognitive science and neuroscience.

I have attempted to integrate these components into a simple but coherent model consistent with the great ape behavioral data, a process that has led me to the claim that apes need a second-order executive tier of functioning. The model thus shares essential features with hierarchical models of executive (cognitive) control, such as that of Koechlin and Summerfield (2007) (as well as that of Shea & Frith, 2019, as a "global workspace" model of metacognition). These models further support the idea of two tiers of executive functioning in apes, as the second-order tier operates on first-order executive processes by, for example, making judgments about the efficiency and reliability of these first-order executive functions, judgments that have so far been observed only in great apes.

The main advantage of this two-tiered psychological architecture is to integrate into one coherent model the functioning of various executive processes—from planning to inhibition to attention monitoring to working memory to metacognition—that have mostly been studied in isolation from one another in both humans and animals. In my conceptualization, each executive tier is itself a feedback control system that attends to processes "below" it and attempts to self-regulate these processes with the goal of making better decisions. An integrated model of this type also enables us to see clearly the intimate connections between decision-making and executive (cognitive) control, for example, in the close ties between go-no-go decisions and global inhibition, on the one hand, and between either-or decisions and more proactive processes of inhibitory control, on the other, as well as between great apes' reflective decisions and metacognitive monitoring.

In any case, definitions and models aside, it is clear empirically that great apes are agentive in ways that are very similar to humans. They perform similarly in many experimental tasks (though often more like human children), including reflecting on and executively controlling their ongoing psychological, and even executive, processes. An obvious but nevertheless profound conclusion, therefore, is that these processes cannot have as their evolutionary or ontogenetic origin anything deriving from uniquely human forms of culture, intentional instruction, or language. Rather, they constitute an evolved system common to all great apes, enabling individuals to make effective and efficient—indeed, reflective and rational—behavioral decisions.

The social medium . . . setting up conditions which stimulate certain visible and tangible ways of acting is the first step. Making the individual a sharer or partner in the associated activity so that he feels its success as his success, its failure as his failure, is the completing step.

—John Dewey, *Democracy and Education*

If great apes are already rational agents, what further form of agency could possibly be open to humans that would account for the many species-unique products and ways of life—involving complex technologies, complex symbol systems, and complex cultural institutions—that enable them to completely dominate the large-mammal niche on planet Earth?

The answer involves one of evolution's oldest tricks, just on a new level. During the past three billion years of life on the planet, a handful of major transitions have occurred in the organization of life-forms, for example, the emergence of chromosomes, the emergence of multicellular organisms, and the emergence of sexual reproduction. In each case the transition occurred in the same basic way: previously independent entities came together to act as a single unit (Maynard-Smith & Szathmáry, 1995). The emergence of humans and their domination of other mammals fit this same general pattern: individuals came together to form socially shared agencies—socially constituted feedback control systems—that could pursue shared goals that no individual could attain on its own.

This happened in two evolutionary steps, the outcomes of which still structure human behavior and psychology today. The initial step was early human individuals (before the emergence of *Homo sapiens*) coming to collaborate with one another in face-to-face interactions to pursue collaborative goals, especially in the context of foraging. Early human individuals formed with other individuals a joint agency. The second step was modern humans (early *Homo sapiens sapiens*, before agriculture and civilization) coming to form distinct cultural groups, each pursuing its own collective goals with its own cultural practices. Modern human individuals formed with others in their cultural group a collective agency. And we know that both of these are true agencies because in both cases a new, socially constituted form of self-regulation—normative self-regulation—obliges individuals to direct and control their actions not just individually but also to comport with the normative standards of the shared agency in which they are participating. Individuals acting in shared agencies are socially normative agents.

Socially shared agencies challenged early humans with many new unpredictabilities, requiring many new psychological adaptations. (Imagine everything we would have to do to our leaf vacuum machine if it had to coordinate with other such machines to collect leaves collaboratively.) What could be riskier and more uncertain than forgoing pursuit of one's own individual goal to try to align and coordinate toward a common goal with a partner who has her own individual goals and values? And the risks and uncertainties are only magnified when the "partner" is an entire cultural group. Making these new socially constituted forms of agency work thus required ancestral humans to develop both new skills of social coordination and new social motivations. And these complex, socially shared agencies have worked spectacularly well for humans—at least so far—leading to a highly successful species with all kinds of new individual skills and motivations, not to mention all kinds of group-level achievements.

Early Human Joint Agency in Collaboration

Chimpanzees and other great apes often forage in the company of others, but the process of acquiring and consuming the food is fundamentally individual. In a typical situation, a handful of chimpanzees travel until they find a fruiting tree, at which point it's everyone for herself. Some groups of chimpanzees also engage opportunistically in the group hunting of monkeys, but this is essentially the same process as that of social carnivores such as lions and hyenas: each individual tries to capture the monkey for itself and, in so doing, takes into account both the monkey's actions and the likely actions of the other chimpanzees. From a psychological point of view, each individual is using the others in the hunt as "social tools" for its own ends.

Humans split off from chimpanzees and other great apes around six million years ago, and then, sometime after one million years ago, began to forage collaboratively. This period was marked by a great expansion of terrestrial monkeys, like baboons, who might have outcompeted humans for their normal fruits and vegetation, pushing them into a new foraging niche. A beginning might have been scavenging meat from carcasses, which would likely have required a kind of coalition of individuals to frighten off other animals interested in the same food. But at some point, early humans began to collaborate more actively in hunting large game and procuring some plant foods, typically in mutualistic stag hunt type situations in which both individuals could expect to benefit from the collaboration if they could somehow manage to coordinate their efforts. This pattern is especially clear in early humans of about four hundred thousand years ago—the common ancestor to Neanderthals and modern humans, *Homo heidelbergensis*—who engaged systematically in the collaborative hunting of large game (Stiner, 2013).

In a stag hunt situation, individuals must collaborate with others to benefit, and the benefits are greater than those of any solo alternatives (which must be forsaken or at least risked). As early humans began obtaining the majority of their food via such collaboration, it became obligate, so that individuals became dependent on one another—they became interdependent—in especially immediate and urgent ways (Tomasello et al., 2012). Another dimension of this interdependence was partner choice. Individuals who were not skilled at collaboration—for example, were unable to communicate effectively—were not chosen as partners. Similarly, individuals who were not cooperatively motivated—for example, tried to hog all the spoils—were also shunned as partners. The upshot was that there was extremely strong social selection for cooperatively competent and motivated individuals (Baumard et al., 2013).

Early humans adapted their skills of great ape rational agency to the challenges of collaborative foraging—that is, the challenges presented by unpredictable partners with their own individual agendas—by developing the capacity to form a joint agency with a rational partner. This required three sets of adaptations not possessed by great apes, supported by a brain about double the size of an ape's brain. (Indeed, in the analysis of

Figure 6.1
Imagined early humans about four hundred thousand years ago.

González-Forero and Gardner [2018], about 60 percent of the brain growth characteristic of early humans during this period was concerned with adaptations for cooperative interactions.) First, humans needed to be able to form with one another a joint goal superseding their individual goals, which required both cognitive and motivational adaptations. Second, they needed to coordinate their individual roles in the collaborative activity, which required new forms of perspective taking and cooperative communication. And third, they needed to have ways of keeping everything on track in the collaboration even in the face of unanticipated exigencies—by working together to cognitively control the collaboration from one or another executive level—which required new mechanisms of social self-regulation. I explicate each of these three sets of adaptations in the three sections that follow.

Forming a Joint Goal

Stag hunt situations present individual foragers with a choice situation full of uncertainties. Specifically, as each individual is gathering low-value foods (hares), a high-value food (stag), whose capture requires collaboration, appears in the distance. Each individual would like to go for the stag, but only if the other individual goes for it as well. If each waits for the other to make the first move, the result is paralysis. How do the individuals reduce the uncertainties of going for the stag?

The answer is cooperative communication, as illustrated in comparative experiments modeling the stag hunt situation. In one study, each of two chimpanzees is gathering for itself low-value rewards, when a high-value reward requiring their collaboration suddenly appears. The result is that one chimpanzee takes off for the stag—heedlessly, as it were—and the other then sees an opportunity and follows. They make no attempt to communicate about the decision at the outset, exposing the first chimpanzee to a significant risk that she will end up stranded on her own at the stag. In contrast, young human children attempt to mitigate their risks ahead of time by making a joint decision to pursue the stag together; for example, one child points to the stag excitedly, or one child entreats the other to follow her verbally, or else they both look communicatively back and forth to the stag and to each other in ways that make clear that they both know together that a stag has appeared (Duguid et al., 2014; Siposova et al., 2018). Such communicative acts reduce risk by establishing in the cooperative cognition, or common ground, of the participants that they both know together that they both want to pursue the stag together: they have formed a joint agency to pursue a joint goal.

Having formed a joint goal, pursuing it successfully requires individuals to jointly attend to relevant situations along the way, as either obstacle or opportunity (analogous to individual agents attending to relevant situations). Thus, if two early humans were collaboratively hunting for game, an antelope at a watering hole would present a relevant opportunity to which they would naturally jointly attend (so as to devise a joint plan of action). A deep ditch between them and the antelope would present a relevant obstacle to which, again, they would naturally jointly attend (so as to devise a joint plan around the obstacle). Young human children engage in joint attention with adults from about their first birthdays, as established both by their coordinated looking patterns with a partner to an external situation and by their active attempts to share attention via pointing and other communicative acts (Carpenter et al., 1998). In contrast, despite many experimental attempts, chimpanzees have never been observed to engage in humanlike joint attention with either a human or a conspecific (even maternal) partner (Tomonaga et al., 2004; Tomasello & Carpenter, 2005; Wolf & Tomasello, 2020a).

Motivationally, the key to forming a joint goal is each partner's assurance to the other that she will play her role in a cooperative spirit: each will subordinate, to some degree, her own individual interests to that of the joint agency, assuming that the other does so as well. This process also requires cooperative communication in the form of a mutually reassuring joint commitment to collaborate (Gilbert, 2014), whose effects may be seen in two lines of comparative experiments. First, if a chimpanzee is collaborating with a partner and unexpectedly gets her reward first, she simply takes it and runs (Greenberg et al., 2010); in contrast, if a young human child gets her reward first, she nevertheless stays committed to the collaboration throughout, delaying cashing in her own reward until her partner gets hers as well (Hamann et al., 2012; see also Kachel & Tomasello, 2019). Second, when a collaboration has reached its joint goal, a dominant chimpanzee will, if possible, hog all the spoils and exclude her partner, which means that the pair is unlikely to continue collaborating (Melis, Hare, & Tomasello, 2006); in contrast, when a young human child obtains rewards collaboratively with a partner, she almost always divides them equally, which encourages continuing collaboration (Warneken et al., 2011; see also Hamann et al., 2011). The joint commitment that children, but not great apes, make and respect is aimed at reducing the uncertainties of collaboration for both participants.

The hypothesis is thus that human children's species-unique ability to form a joint agency to pursue a joint goal, employing skills of joint attention and cooperative communication, reflects adaptations that occurred several hundred thousand years ago in early humans. Likewise, human children's species-unique motivation to make joint commitments with one another—and so to reduce risks by mutually subordinating their individual goals to those of the joint agency—reflects adaptations from the same evolutionary period. The foundation stone of uniquely human agency is individuals' ability and propensity to form with others a joint goal, thereby creating an evolutionarily unique, socially constituted feedback control system.

Coordinating Roles

Joint goals and joint attention created for early human collaborators a kind of shared world in which to operate. But to collaborate effectively, they each had to recognize at the same time the individual goals and perspective of their partner in her individual role. For example, in hunting an antelope, one partner might have played the role of chaser, which had one goal and perspective on the shared situation, while the other partner played the role of spearer, which had another goal and perspective on the shared situation. Coordinating these roles skillfully toward a common end required, first, that both partners understood each other's role and perspective, and second, beyond this if possible, that both partners facilitated each other's role and perspective via acts of cooperative communication.

When chimpanzees are given the opportunity to play a novel role in a collaborative activity, they learn and become proficient in it in the same way regardless of whether they have previously played the opposite role in that activity. In contrast, if young children have played the opposite role first, they know immediately how to play the novel role, presumably because whenever they collaborate, they understand not only their own role but also that of their partner (Fletcher et al., 2012). This species difference is also apparent when individuals are forced, in the midst of a collaborative activity, to reverse roles, which human children, but not chimpanzees, readily do (Tomasello & Carpenter, 2005). Young children in a joint activity thus understand what their partner is trying to do in her role (her goal), and thus they can identify the situations that are relevant to that role as both obstacles and opportunities (from her perspective), which enables them to coordinate both roles mentally.

When a collaborating individual notices a situation relevant to the joint goal that her partner has not yet noticed, it would be helpful if she could draw her partner's attention to that situation. Early humans thus evolved a new form of communication—cooperative communication (also used to

form joint goals and commitments; see previous section)—in which collaborative partners informed one another of things helpfully so as to facilitate their joint success (Tomasello, 2008). The first instantiations were the
species-unique gestures of pointing and pantomiming. To communicate
effectively using such natural gestures, individuals had to take each other's
perspective: I see that you are not attending to something that I am; or
conversely, I try to discover what you are apparently attending to that I am
not (because you are gesturing to me). Chimpanzees do not communicate
in this way. In chimpanzee group hunting, an individual may be excited
about an approaching monkey and scream, from which other individuals make inferences; but the screamer does not intend this effect. Indeed,
one of the most striking observations about chimpanzee collaboration in
experiments is that they do little, if anything, to actively communicate
with their partner, even when it would be easy and useful to do so (Melis
et al., 2009). Chimpanzees are not attempting to influence their partner's
perspective in the direction of joint attention because they do not operate
with the notions of perspective and joint attention in the first place.

Although natural gestures are obviously not as powerful as a language,
early humans used pointing and pantomiming to collaboratively plan and
make decisions together toward joint goals, as well as to adjust to unanticipated situations along the way. For example, drawing on their common
ground of shared experiences, two individuals might decide to go fishing
together as one of them points to the river and the other agrees, or they
might decide to hunt antelopes as one of them mimes using a spear and
the other agrees. During the collaborative activity itself, they might point
and pantomime to coordinate their respective roles in the moment, for
example, by pointing to indicate things like "you go here, and I'll go there"
or "spring the trap now." Cooperative communication of this type both
relies on and facilitates unique cognitive skills of mental coordination: the
individuals must simulate one another's perspectives as they attempt to
align perspectives in joint attention to relevant situations (see Tomasello,
2008, for a review of the evidence). Because human children naturally
communicate in this way, they find it trivial to locate hidden food when
someone points to its concealed location; they immediately know that
the pointing individual intends to help them discover where the food is
located. Great apes do not make the same inference in this simple situation
because it involves both assuming the pointer's cooperative motivation

(to help her partner find the food) and a recursive mental coordination—she *intends* for me to *know* that the food is in this bucket—and apes simply do not have such motivations or make such inferences. They do not because they did not evolve to make their living by collaborating with others (Tomasello, 2006).

I thus hypothesize that early humans evolved some species-unique cognitive skills—including, most importantly, perspective taking and cooperative communication—to mentally plan and coordinate joint agencies working toward joint goals. But, just as in the case of individual agencies, in joint agencies things do not always go according to plan, and so there also is a need to cognitively control or self-regulate the process as well.

Collaboratively Self-Regulating the Collaboration

The two main mechanisms by which individuals can procure better collaborative partners are partner choice and partner control. In the obligate collaborative foraging characteristic of early humans, partner choice meant selecting the best partner (and excluding others). In contrast, partner control was not about choosing a partner at the outset, but rather about attempting to make a current partner behave better. The best-known instances of partner control in the animal kingdom occur when individuals punish others for not doing what they want them to. But early humans created a novel form of partner control. After the collaboration was ongoing (and it would be costly to unchoose the current partner), the individual could communicatively protest the partner's behavior: you are not being cooperative, so I am threatening to opt out and leave you on your own. The aggrieved partner thus gives the transgressor a second chance to voluntarily mend her ways before she is partner choiced out of the picture.

In their group hunting of monkeys, chimpanzees do not engage in partner choice; hunts typically begin opportunistically with whoever is in the vicinity. Nor do chimpanzees engage in partner control: they do not protest the behavior of wayward participants in the midst of group hunting or in any other contexts. In contrast, when three-year-old children have made a joint commitment to collaborate, and their partner does not play her role in the ideal way, they protest to her (which they do not do if she is ignorant of how to behave in this role; Kachel et al., 2018). Children also protest if their partner to a joint commitment just up and leaves without an excuse or apology (Kachel et al., 2019). At the end of the collaboration, chimpanzees do not punish or exclude free riders who attempt to grab some of the spoils without having participated (Boesch, 1994; Melis et al., 2011; John et al., 2019), whereas young children actively protest both against free riders (Melis et al., 2013) and against partners who attempt to take more than their share (Rakoczy et al., 2016). The evolutionary hypothesis is thus that early humans attempted to control their collaborative partners with communicative protest.

Of crucial importance to the appropriate interpretation of this behavior, human children, and so presumably early humans, express their protest in normative terms of what one *must* do, or *ought* to do, or *should* do, or *has to* do: "You have to do this!" And such normative protest often works to shift the behavior of the transgressor in a more cooperative direction. The process may thus be seen as a kind of collaborative self-regulation. Importantly, although the protest is emanating from one partner, it is seen by both partners as coming from "we." The protest is not just that I don't like it, but rather that one *should* not do this but *ought* to do that. The normative stance is thus not a personal preference but a harking back to the joint commitment that "we" proceed in a cooperative spirit. From the outset, each signatory to the joint commitment entitles the partner to call a transgressor out for bad conduct, meaning that the transgressor will agree that she deserves to be censured for her uncooperative actions (the protest is agreed to be legitimate). Normative protest for inappropriate conduct thus implies shared normative standards that each partner may invoke to indicate a shared judgment that an action is detrimental to the joint agency, and the partner is obligated to respond appropriately. Normative protest thus entails not one partner controlling the other but the joint agency controlling or self-regulating itself.

Because of their sensitivity to all of this, young children, and so presumably early humans, engage in some peculiar preemptive behaviors that are unique to the species. For example, it is almost certainly unique to humans that when a partner to a collaboration needs to leave for some reason, she will "take leave" or make an excuse or even ask permission, explicitly recognizing that she may not legitimately just break off without reason (Gräfenhain et al., 2009). The initial agreement in the form of joint commitment thus makes partners feel responsible to each other to live up to their shared common-ground expectations, what "we" each ought to do according to "our" standards. If an individual does not, she should apologize to her partner or, if all else fails, feel guilty and attempt to repair the damage (Vaish et al., 2016). Overall, Tomasello (2020) argues and provides evidence that joint commitments engender in young children a sense of responsibility to their collaborative partner, and this normative sense of responsibility is the "glue," as it were, that keeps the joint agency intact even in the face of temptations to defect.

Thus early human individuals not only collaborated to achieve joint goals but also collaborated to self-regulate the collaboration. Each partner operating within a joint agent "we" must behave cooperatively to pursue the joint goal for their joint benefit, or else that joint agent "we"—as represented legitimately by either partner—will act to bring the wayward individual back into line. In the context of the collaboration, "we" always has the last word. The result was a kind of "we > me" sociomoral self-regulation, such that each partner internalized a responsibility to play her role in the joint agency—the individual used the joint agency to self-regulate her individual behavior—in a way that comported with their common-ground normative standards.

Cooperative Rationality and Its Experiential Niche

Early humans thus for the first time began putting their rational heads together with a partner to form a joint agent to pursue a joint goal together. These collaborative activities were dual level in the sense that they comprised a shared level of joint goals and joint attention, on the one hand, and an individual level of individual roles and individual perspectives, on the other. We might think of these as two *modes* of agency.

Figure 6.2 shows the mode of joint agency in the middle, in a box with borders. It corresponds to the three tiers of agentive functioning—operational, executive, and reflective—from great apes, but cooperativized, that is, with the word *joint* in front of each component: joint goal, joint attention, joint decision, joint action. At the bottom of the figure are the partners' individual role agencies, in which each acts to carry out her role in the collaboration. Role agents—whose internal workings are not depicted in the figure—have their goals set by the joint planning and decisions of the joint agent "we"; their individual plans and intentions in their roles are subordinated to the joint plans and intentions of the partnership in a manner that enables them to be meshed appropriately (Bratman, 2014). But another mode of agency is actually at work here as well: the individual

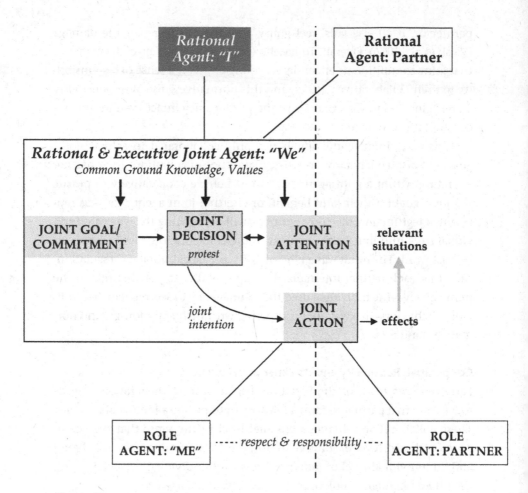

Figure 6.2
Joint agency as cooperativized rational agency. This comprises three separable modes of agency, represented here as the rational agent "I," the joint agent "we," and the role agent "me." Everything in this diagram is highly simplified. Further explanation in the text.

rational agent (at the top of the diagram, whose internal workings are also not depicted) that decides whether to collaborate or make a joint commitment in the first place. This rational agent constantly assesses the situation as it unfolds with respect to her own individual interests, such that she can at any time decide to opt out of the collaboration (either with or without permission). Of course, only one human being is involved here; but this human being is acting simultaneously as an individual "I" pursuing her own self-interest, a joint agent "we" pursuing a joint goal with a partner, and a role agent "me" whose behavior is directed by the joint agency. Early human individuals thus operated with what we may call a cooperative rationality—they did what made sense in the context of their collaboratively structured agency—and this required them to juggle simultaneously the operations and co-operations of three distinct but interrelated modes of agency.

Early human cooperative rationality constituted a radically new psychology both cognitively and socially. Cognitively, to mentally coordinate with a collaborative partner, including via cooperative communication, early humans evolved to cognitively represent the world perspectivally: the exact same object or event may be construed as something different depending on the perspective one chooses to take. For example, this stick on the ground might be seen as a potential spear for us to use in our antelope hunt (if we need a weapon), or it might be seen as something that could make noise if stepped on (if we are worried about that), depending on our common-ground understanding of what is relevant to the situation at hand. Since the process of mental coordination in cooperative communication required individuals to take the perspective of others on their own perspective recursively—he *intends* for me to *attend* to that as a potential weapon—early humans came to cognitively represent the world both perspectivally and recursively (see Tomasello, 2014, for a review of relevant evidence). Great apes have not evolved recursively perspectival representations because they have not evolved to mentally coordinate with others in joint agencies.

Socially, *co*-operating with others in this way also required new sociomoral attitudes and emotions (see Tomasello, 2016, for a review of relevant evidence). First, since collaborative partners see each other as having equal status, collaborative partners *respect* each other, as evidenced by the fact that they treat them *fairly* and *entitle* partners to rebuke them *legitimately* for uncooperative behavior. Second, if one partner does not treat another fairly, she *resents* being treated as less than equal and therefore protests normatively. Third, the result is that partners feel a *responsibility* to treat each other cooperatively in the way that they *deserve* to be treated as specified by their joint commitment at the outset. And finally, if a collaborative partner succumbs to selfish motives at some point, she must make an *excuse* or *apologize* to her partner or else suffer feelings of *guilt*. All these italicized terms depend, in one way or another, on the shared normative standards by which "we" evaluate and self-regulate "your" and "my" actions as coequal partners. Great apes and other mammals have not evolved these normative attitudes and emotions—they certainly do not make excuses or apologize for their own bad behavior—again because they have not evolved to *co*-operate with others in joint agencies with normative self-regulation.

It is also worth pointing out, at least in passing, that human social relationships seem to rest to a very large degree, and in a way seemingly not present in great apes, on shared experiences and common ground. Thus early human individuals likely felt closest to those with whom they shared the most experiences—best friends are those with whom one shares the most—where "sharing" means experiences derived through joint attention and common ground, prototypically in collaborative activities. Clearly, the sharing of experience at work here is recursive in that each partner knows that the other also is sharing experience with her, and knows that she knows this as well, so that everything is reciprocal. And so, arguably, human social relationships in general—including the normative dimension of what we owe to friends as a function of our closeness with them—derive from the fundamentally cooperative nature of human social relations (see Wolf & Tomasello, 2020a, 2020b, for an experimental demonstration in young children).

As I have argued at other steps in my story, new agentive organization creates for individuals a new experiential niche. Reptiles come to experience situations of obstacle and opportunity; mammals come to consciously experience their own operational level of functioning; and great apes come to experience their own executive decision-making and cognitive control from a reflective tier of operation, which serves as the basis for apes' understanding of causal and intentional relations in their physical and social worlds. Early humans came to live in a social/cooperative experiential niche, structured by the shared worlds and recursive perspectives created by collaboration, joint attention, and common ground, and motivated by the partners' sense of respect and responsibility toward one another. Shared worlds experienced via recursive perspectives among mutually respectful and responsible cooperative agents: this is the new experiential niche inhabited by early humans.

Modern Human Collective Agency in Cultural Groups

Early humans collaborated in pairs, but they lived in larger, loosely structured social groups. That worked well for several hundred thousand years, but then, about 150,000 years ago, it began to work less well, and the reasons were mainly demographic. The problem was that these groups were so successful that their populations kept growing, which meant ever more encounters and possibly conflicts with other groups over territory and resources. By the end of this period, we observe the emergence of distinct cultural groups that distinguished themselves from one another—even from neighboring groups—by operating with distinct sets of artifacts, which clearly required different knowledge and cultural practices.

For modern human individuals to survive and thrive, they needed to stay in their group, and if they were threatened by other groups, they needed to band together. The new psychological mechanisms for banding together, not available to other apes, were the skills and motivations for joint intentionality and agency bequeathed to modern humans by their early human ancestors. Modern humans scaled up these skills and motivations beyond the foraging pair to the social group at large, constituting new skills of collective intentionality. Social life within a modern human group thus gradually became one big collaborative activity, with the collective goal of group prosperity in the face of competition from neighboring groups. As one important example of such group-level functioning, there emerged with modern humans a new form of collective foraging known as central-place foraging. When smaller parties obtained large packets of food, such as large game, they brought them back to a central location and shared the bounty with the entire social group. In general, the group shared an ethos of trust and loyalty, similar to that within their nuclear families.

As groups grew, they began to fractionate (perhaps because human individuals can only cooperate effectively on a personal level in groups of 150 or so individuals: Dunbar's number). This created so-called tribal societies, or cultures, which comprised multiple social bands, living apart, but still with some sense of the larger cultural entity. If they did not scale up their collaborative skills and loyalty to the cultural group at large, they risked being outcompeted by other groups. Using historical evidence, Samuel Bowles and Herbert Gintis (2011) have argued that competition between modern human groups was in fact the main driver of ever greater degrees of in-group cooperation. Peter Turchin (2016) has documented empirically, with several different kinds of evidence, that groups structured by a greater degree of cohesion, solidarity, and group commitment tend to do better in intergroup competition than groups not so structured. It is thus competition among groups that creates the ecological challenges pushing small, loosely formed groups of early humans into the collective agencies of modern human cultural groups. The process was so strong that it even made cultural groups into coherent units of natural selection. Robert Boyd and Peter Richerson (2005) have argued that cultures with "strong" conventions, norms, and institutions supported by individual group members' loyalty and conformity form units of selection that tend to survive and persist, whereas cultures with "weak" conventions, norms, and institutions not supported by individual group members' loyalty and conformity fractionate and perish.

The process by which modern human children became skillful and motivated members of their cultural group involved a greatly extended ontogeny of dependence. Whereas juvenile chimpanzees become independent of adults for food immediately upon weaning at around four years of age, modern human children were (and are) dependent on their parents and other adults for food for another decade, well into adolescence. During this extended period of dependence, individuals learned—indeed, were expected by adults to learn—the ways of the group and how best to support these ways. Modern humans' very slow ontogenies are illustrated by the very slow pace of brain development as compared with that of other apes (and likely early humans; Gunz et al., 2019). Chimpanzee brains reach 90 percent of adult size already by two years of age, whereas modern human brains do not reach that mark until eight years (Coqueugniot et al., 2004). In the end, modern human brains are three times larger than those of other great apes, with an expanded prefrontal cortex (the main seat of executive functioning) and insula (the main seat of social emotions), and are structured by unique types of neurons with more complex dendritic structures

Figure 6.3
Imagined modern humans about one hundred thousand years ago.

(Kaas, 2013; Donahue et al., 2018). Consistent with this analysis, González-Forero and Gardner (2018) find that much of the brain growth characteristic of modern humans during this period was concerned with adaptations for cooperative interactions and cultural learning.

Forming Collective Goals

Modern human cultural groups thus became collective agencies that pursued goals and made decisions as a single body. Collective goals concerned matters such as the destination of group travel, the location of a home base, the preparations for group defense, the division of resources, and the division of labor in tasks such as collective childcare. Collective decisions about such things took place through discussion resulting in consensus. Some individuals' voices might be louder than others, but in virtually all small, informal human groups, when a large majority inclines in a certain direction, the rest tend to go along. And if one individual tries to take too much power and impose his will, the rest of the group pushes him aside or worse (Boehm, 1999).

This new collective way of doing things resulted in two major changes in individual psychology. First, because individuals were almost totally dependent on the group for survival, they became ever more concerned with its well-being, expressing their loyalty in various ways. Combined with a mistrust of outside groups, the result was a distinctive in-group/out-group psychology. In-group favoritism accompanied by out-group mistrust is one of the most well-documented phenomena in all of social psychology, and it emerges ontogenetically already in childhood. Thus children are more inclined to help in-group than out-group others; they are prone to share more with in-group than out-group others; and they care more about their reputation with in-group than out-group others. Reciprocally, young children also favor their in-group compatriots who express loyalty to the group (and mistrust of out-groups). In a direct test of the evolutionary hypothesis that group competition spurs within-group cooperation, groups of young children are more cooperative within their in-group if they are competing with an out-group (see Dunham, 2018, for a review of all this literature).

Second, because individuals needed to be able to recognize in-group members and, just as important, to be recognized by them as a member of the in-group, they began to conform to one another's practices as a way of expressing their group identity (which worked even with in-group strangers from other social bands within the cultural group). If everyone in the group has a tendency to conform, then people who talk like me, dress like me, and eat the same foods as me are most likely members of my cultural group, even if I have never met them before. And, once again, conformity to the in-group is one of the most well-documented phenomena in all of social psychology. Thus, when young children witness in-group peers making clearly incorrect judgments, they often conform and follow suit nonetheless, especially if those peers are watching (Haun & Tomasello, 2011). Further, even after they have already solved a problem successfully, if young children see in-group peers solve it in a different way, they often switch to follow the crowd (whereas chimpanzees stick with what has worked for them individually in the past; Haun & Tomasello, 2014). And slightly older children even tend toward what has been called overimitation—copying causally irrelevant aspects of an action—as they take whatever adults do to be an expression of the normative way that "we" do things (Keupp et al., 2013). In this way, modern human individuals came to actively conform to the group's ways—including even arbitrary conventions and religious rituals as "costly signals" of group membership—to actively display their group identity.

The power of group identity on individual behavior is spectacularly apparent in so-called minimal group experiments with young children (also apparent in experiments with adults). When preschool children are simply told that they are members of "the green group" (and given a green scarf to wear), they are immediately more cooperative with others in the green group—they help them more, share with them more, trust them more, and so on—than they are with children in other-colored groups (again see Dunham, 2018, for a review). Without an evolutionary basis of some kind, it is basically incomprehensible that someone should feel solidarity with arbitrary others based simply on commonalities of appearance. Such appearance-based solidarity in modern humans meant that now there were two bases for solidarity with others: collaboration (inherited from early humans) and similarity. The importance of similarity-based solidarity has led Daniel Haun and Harriet Over (2015) to propose homophily—the tendency to affiliate, favor, and bond with similar others—as the psychological basis of human culture.

Modern humans' collective agency was thus made possible by individuals evolving a group-minded concern for the culture's goals and welfare, and a propensity to conform to the group's ways of doing things. Noncooperative individuals were excluded from group benefits, thus creating a kind of group-level social selection that Brian Hare and colleagues (2012) have called self-domestication, since it leads to the social selection of individuals who are tolerant and cooperative in the group. The result was a collectively constituted agency that pursued collective goals.

Coordinating Societal Roles

From the point of view of individuals collaborating with one another within the cultural group, the challenge for modern humans was coordination with in-group strangers. As cultural groups became too large for individuals to know and have personal common ground with all other members (in the manner of early human joint agency), collaboration and communication suffered. The solution was that within a cultural group there arose a new kind of shared experience based on *cultural* common ground, that is, based not on individuals' personal experience with one another but on a commonality of experience assumed simply on the basis of assumed group membership. Such cultural common ground assumed that common experiences in the culture led to common skills and practices, on which individuals who wished to collaborate could jointly rely (i.e., commonalities helped not only with group identification but also with collaboration).

Initially, individuals could learn to participate in the group's conventional cultural practices simply by observing and culturally learning from others. For example, a modern human child might observe experts engaging in the practice of net fishing and conform to their ways of doing it in each role, since, as we have just seen, children have a strong tendency to conform to in-group others. But imagine further that the child now wishes to net fish with some in-group strangers. Do they know how to net fish? The guiding assumption of modern human individuals was that they did, because everyone who grew up in the cultural group—identified by their language, dress, and so on—knows how to net fish. Thomas Schelling (1960) has argued that common ground of all types—including cultural common ground—requires a kind of recursive mind reading: the individual expects that in-group strangers will conform to the conventional practice, *and* they will expect her to conform, *and* they will expect her to expect them to conform, and so on. It is only with such mutual expectations of conformity that in-group strangers can come to the river and immediately begin net fishing smoothly with one another. As evidenced by contemporary Western children, if a novel in-group adult looks in the general direction of a strange doll and a Santa Claus doll and says, "Oh, I know that one, can you hand it to me?" young children hand over Santa Claus. Although they have never before interacted with this in-group stranger, they assume that she is familiar with Santa Claus, but not with the novel doll (Liebal et al., 2013). Coordinating with in-group strangers to form a joint agency requires a kind of group-minded cognition emanating from the collective agency of the group: mutual expectations in cultural common ground.

Coordination with all kinds of partners was facilitated, of course, by cooperative communication. Communicating with a familiar partner in a mutually known collaborative activity requires fairly simple means, for example, early humans' natural gestures of pointing and pantomiming. But modern humans needed to communicate in the wider context of a cultural group, including with in-group strangers, and this presented new challenges. These challenges were basically coordination problems—the interlocutors needed to coordinate their attention to a common referent— and so the solution, as for all of modern humans' other coordination challenges, was to conventionalize their natural communicative activities into a set of conventionalized cultural practices that everyone knew that everyone knew (Lewis, 1969). Whereas in communication with natural gestures the common ground needed is only about situations in the world and their relevance for a collaborative activity, in conventional communication the communicative conventions themselves must be in cultural common ground. Both partners know that they both know that anyone in this cultural group will coordinate attention in the appropriate way when they use particular pieces of the conventional language (e.g., they will jointly attend to my spear if I say "my spear"). Conventional languages thus facilitated ever more complex and long-range planning and decision-making toward shared goals, with a coordination of individual roles, including with in-group strangers. In addition, conventional languages were used in pedagogy, as adults informed children of facts and skills that should be useful to them based on culturally normative knowledge (Csibra & Gergely, 2009).

Just as early humans' joint agencies had individual roles, so modern humans' joint agencies (within the group's collective agency) had individual roles as well. The difference was that these were roles that everyone knew that everyone knew (in cultural common ground) how to perform. In addition, however, as human cultures evolved, there emerged societal-level roles. Thus a modern human cultural group might have a subset of members who hunted large game, another subset who gathered resources, a specialist who dealt with ill or injured individuals, specialists who made various tools, a group leader, and so on. Those who performed such societal roles had special rights to do certain things, but also special responsibilities to the group. With such division of labor, the cultural group became ever more like a single "superorganism," a fully fledged collective agent whose individual members survive and thrive only if everyone does his or her job.

Collective Self-Regulation via Social Norms

One of the most robust findings in all the social sciences is that cooperation becomes more difficult as group size increases. This happens for many
reasons, including most importantly the diminishing proportional contribution of each individual (my contribution matters less, so why bother)
and the decreasing probability that cheating or free riding will be detected
(I am more anonymous and so may get away with shirking). As cultural
groups become larger, then, various kinds of social dilemmas arise in which
individuals' group-minded motives compete with individual motives of
self-interest. This means that when there are collective goods from which
everyone in the group benefits—for example, common sources of water or
firewood—the temptation can arise for individuals to take what they can
in case others are doing the same, leading to a "tragedy of the commons,"
in which the collective goods are depleted for all. Individual behavior must
be self-regulated by the group—in a kind of group-level "we > me" ordering—or else everyone suffers.

Early human individuals made joint commitments, and they self-
regulated them by entitling each party to call the other out on behalf
of the joint agency (via protest) for transgressions. Now, in the collective
agency of a cultural group with all kinds of cultural conventions and roles
as part of the cultural common ground, there emerged collective expectations for individual behavior, also known as social norms, that served as
self-regulators. All modern human cultural groups had (and have) social
norms, at the least to self-regulate activities in which group-damaging conflicts might occur, for example, in dividing resources or in access to mates
(see Tomasello, 2016, for a review of relevant evidence). Importantly, the
regulatory "we" of a social norm was the cultural agency as a whole: everyone expected everyone in the cultural group to conform to its conventions
and norms (and they expected everyone to expect them to conform as
well). Social norms are about conformity to the group's ways, and if the
group is to function smoothly, everyone *must* be committed both to following those norms and to calling out transgressors for the good of the
group as well.

It is clear why individuals should follow cultural norms—to be accepted
by the group and to avoid sanctions—but it is not as clear why they should
enforce them on others. After all, sanctioning is risky if the transgressor
resists. But even three-year-old children routinely enforce social norms

on others. For example, children tell others that they should not damage someone else's toy or that they should play a game in the conventional manner (see Schmidt & Tomasello, 2012, for a review). Importantly, if the transgressor is an in-group member (as identified by his language), children hold him to a higher standard, presumably because he participates in the cultural common ground of the group and so should know better (Schmidt et al., 2012). Children three years and older—and so, by inference, early modern humans—enforce social norms because they implicitly understand them as the means by which the group regulates itself. As members of the group who care about its fate, children call out in-group members whose nonconformity threatens the group's smooth functioning. Enforcing norms means looking out for the group's welfare by enforcing its collectively understood norms for individual behavior, thus constituting a new collective, group-level form of the basic "we > me" self-regulation of all shared agencies.

Interestingly, young children approaching school age often create, in play situations, their own social norms and then enforce them on others (e.g., Hardecker et al., 2017). This suggests that the force of social norms for school-age children (and so for early modern humans) comes not from any kind of authority but rather from the social agreements that created those norms. Even young children thus understand social norms as group-created, group-level commitments that "we" use to self-regulate "us." And since "we" made them, they are *legitimate* self-regulators of our conduct. Internalization of the process of group-level "we > me" self-regulation thus leads individuals to feel not just a responsibility to a partner but an obligation to the cultural group and its normative standards, and even to feel guilt for transgressions against those standards. "We" collectively self-regulate everyone, including myself.

Normative Rationality and Its Experiential Niche

Just as joint commitments with responsibilities to partners constitute the motivational "glue" of joint agencies, so collective commitments with obligations to the cultural group and its social norms constitute the motivational "glue" of collective agencies. Modern human individuals who internalized the process of collective self-regulation thus became not just rational or cooperative agents but fully normative agents operating with a normative rationality of obligation (Tomasello, 2020). Once again, this meant that modern human individuals were operating simultaneously with three modes of agency: an individual "I" pursuing her own self-interest, a collective agent "we" operating via the group's collective practices and norms, and a role agent "me" performing the duties that we in the cultural group oblige me to. (Depicting this manner of functioning in a diagram would result in the same basic diagram as the depiction of early human cooperative agency and rationality in figure 5.2, except that everything that is "joint" in that figure would now be "collective" in the new figure, i.e., based not on a partner but on the group. I will therefore not draw it all out explicitly.)

As in every step of our story, this new form of agentive organization created for modern humans a new experiential niche. As a species of great ape, modern humans perceived and understood their physical and social worlds in terms of underlying causal and intentional forces. As descendants of earlier humans, modern humans perceived and understood reality in terms of different possible perspectives on it, and also in terms of newly normative social attitudes, like responsibilities, that bound individuals to their collaborative partners. But as they evolved into fully cultural beings, modern humans came to perceive and understand the world not just in terms of individual perspectives on things but in terms of the objective situation that was independent of any individual perspective. And they came to understand their group mates not just in terms of their responsibilities to one another but also in terms of their obligations to uphold the collective normative standards agreed to by everyone in the group. Modern humans came to inhabit an objective-normative world.

The key cognitive advance creating this objective-normative world is the ability to distinguish between subjective perspectives or beliefs, on the one hand, and the objective situation or reality, on the other. Great apes do not make this distinction: they take the world as it appears to them, and act accordingly. They can discern what a conspecific is perceiving, but they do not contrast his perspective with their own perspective on the situation, much less with the objective situation (because they do not understand perspectives as contrasting views of the same thing). Modern human children come to make the distinction between subjective and objective sometime between four and five years of age, as they come to a full understanding of beliefs (including false beliefs) as mental states that may or may not match the objective situation. The process is not just "reading the mind" of another, as apes already do, but rather mentally coordinating with others in a way that requires the comparison of different perspectives or beliefs on one and the same reality.

This process is made possible by evolved capacities for, and ontogenetic participation in, shared agencies. This is because shared agencies comprise both shared experience on a common focus along with different perspectives. Thus empirical studies show that children's coming to understand beliefs and their relation to objective reality begins in their earliest cooperative communication as they attempt to mentally coordinate with a partner's perspective toward an object of joint attention, which then scales up to perspective-shifting discourse in a conventional language (see Tomasello, 2018). My hypothesis is thus that as early modern humans began to express and exchange perspectives with one another in the medium of a conventional language, reflected on by two tiers of executive function, they constructed for themselves the distinction between the subjective beliefs of individuals and an objective reality. We may envision the ontogeny of this constructive process as a kind of "representational redescription" (Karmiloff-Smith, 1992), in which the individual generalizes and transforms, on its reflective tier, the fact of indefinitely many perspectives on the same situation into something like a perspectiveless, that is, objective, perspective (see Nagel, 1986).

The result was modern human individuals who understood that an agent's view of a situation may be correct or incorrect, depending on whether the view matches the objective situation. They then instructed their young with this in mind. Thus, as Csibra and Gergely (2009) have argued, natural pedagogy takes an objective stance on things from both roles: teachers are not giving their personal opinions about things, but rather transmitting objective facts as enshrined in their cultural knowledge; and children are predisposed to understand such instruction to be about the objective world. The same applies to social norms: adults are not giving to children their personal preferences, but rather informing them about objectively correct ways to do things; and again children are predisposed to understand social norms transmitted in this way to be about the objective world (and thus objectively valid; for an experimental demonstration with young children, see Li et al., 2021). Normative rationality thus means adapting one's individual agency to "objective" facts and values as they inhere in collective cultural experiences.

Modern human agency thus operates in a world of objective facts and objective moral values. And, in one of the most curious phenomena of the natural world, individuals extend this objectivity to their social-institutional worlds to create what John Searle (1995) calls social facts or institutional reality. Social facts and institutional reality comprise real and powerful entities such as husbands, wives, and parents and their respective rights and responsibilities (created by recognition of the cultural ritual of a marriage ceremony); leaders or chiefs and their rights and responsibilities (created by group consensus and sometimes a ceremony); medicine men and their rights and responsibilities; and so forth. They can also turn otherwise ordinary objects, such as shells or pieces of paper, into culturally potent entities such as money. The phenomenon is that a normal person or object acquires a new status based solely on the deontic powers she or it is collectively given by the group via some form of agreement, and that agreement is objectified and so becomes part of external reality. Clever as they are, chimpanzees (and human infants) cannot act meaningfully in modern humans' social-institutional world—they do not recognize chiefs and parents and money, with their respective deontic powers—because they do not have the capability of conferring new normative statuses on otherwise ordinary persons and objects by "agreement." The result, in the evocative description of Yuval Harari (2015), is as follows:

> Over the centuries, we have constructed . . . a reality made of fictional entities, like nations, like gods, like money, like corporations. And what is amazing is that as history unfolded, this fictional reality became more and more powerful so that today, the most powerful forces in the world are these fictional entities. Today, the very survival of rivers and trees and lions and elephants depends on the decisions and wishes of fictional entities, like the United States, like Google, like the World Bank—entities that exist only in our [collective] imaginations.

The word *fictional* in this context does not mean "not real," as these entities are all too real. It simply means "brought into existence by human consensus."

And so, as modern human children came to maturity in their culturally structured experiential niches, they were under constant pressure from the culture and its pedagogy and social norms to believe and do the objectively "right" things. Once humans began operating with objective cognitive representations and objective moral values, these pressures transformed them into what we may call normatively rational agents, who thought about things and did things in the right way for the right reasons. This enabled them to become fully competent participants with their compatriots in a normatively structured cultural world. Normatively structured cultural worlds channel human ontogeny so powerfully that some cultural anthropologists have even claimed that the cultural/normative dimension of agency supersedes individual rational agency. Even if a person decides to act selfishly for her own interests, this decision is made by one who has come to maturity within a particular cultural context (e.g., individualistic) that has shaped her identity and values, and thus her every decision (e.g., Geertz, 1973).

The Complexities of Human Agency

Most of the unique psychological capacities of the human species result, in one way or another, from adaptations geared for participation in either a joint or a collective agency. Through participation in such agencies, humans evolved special skills for (i) mentally coordinating with others in the context of shared activities, leading to perspectival and recursive, and ultimately objective, cognitive representations; and (ii) relating to others cooperatively within those same activities, leading to normative values of the objectively right and wrong ways to do things. Individuals who self-regulate their thoughts and actions using "objective" normative standards are thereby normative agents, very likely characterized by a new form of socially perspectivized consciousness, what we might call self-consciousness.

Great apes can experience a motivational conflict. They may want a piece of fruit, which suggests going for it, but a dominant individual is nearby, which suggests refraining. But this is just the basic instrumental decision-making characteristic of all intentional and rational animals. The claim here is that humans actually operate with different agencies, each with its own goals, which may potentially conflict with one another (Rochat, 2021). For example, if on a collaborative hunt I capture a small mammal, my hunger directs my rational agency to eat it; my sense of commitment to my partner directs my joint agency (and its embedded role agency) to beckon to my partner to come and share; and my sense of obligation to the cultural group's social norms directs my normative agency (and its embedded societal role agency) to bring my catch back to camp to share all around. This is not the garden-variety decision-making of other intentional and rational agents, because each of these agencies has a different mechanism of self-regulation: individual self-regulation, joint self-regulation via normative protest, and collective self-regulation via the cultural group's social norms. These represent three different feedback control systems, each with its own multiple tiers of organization.

The philosophical question of how we should think about socially shared agencies—are they really agents?—has proponents on both sides. Some theorists have no problem with them, and indeed, some biologists believe it is most accurate to think of even an ant colony or a beehive as constituting a single "superorganism" (Wilson & Wilson, 2007). Some philosophers feel similarly about the collective agency of human institutions such as corporations and governments (List & Pettit, 2011). Other theorists follow the spirit of Thatcherism to the effect that society is nothing more than some number of individuals interacting. But surely both views are in some sense correct. When the individual makes a joint commitment with a partner to participate in a collaborative activity or a collective commitment to the group's cultural goals and norms, a new mechanism of goal pursuit, accompanied by joint or collective self-regulation, is thereby created. But in doing so, participating individuals do not lose their individual rational agencies; no matter how difficult or counterproductive it may seem, the individual rational agent may, in one way or another, always opt out. Modern humans are individual rational agents who sometimes (though not always) subordinate their individual agencies to various shared agencies when doing so is either instrumentally or normatively appropriate, with ontogeny in a culture helping to shape those judgments of appropriateness.

Evidence for this view comes from the fact that humans experience normative—that is, moral—dilemmas about what is the best thing to do. Thus an individual may be committed to keeping her promise to a partner (to share food at the end of a hunt), but then this might conflict with the culture's social norms (to bring back large packets of food to the camp); and if I am generous and let my partner have all the food, what will happen to my family? Once an individual enters a shared agency, other "voices" may tell her what she ought to think or do. And unlike the case of the great ape deciding whether or not to pursue some food, there is not a single best answer. Whereas the ape makes a single cost-benefit computation that, if performed accurately, is instrumentally decisive, when I, as human, try to decide whether it is better to be generous toward my partner or to break my promise to my partner or to neglect my obligation to my culture, there is no single best answer. That is why humans experience genuine moral dilemmas, and that is one reason why we should conceptualize the situation not as normal decision-making but as conflicts within an individual among different agencies with different goals and values.

We are left with a puzzle about human normative agency. On the one hand, human individuals would seem to have more choice, more "free will," than other creatures (see List, 2019, for the argument that "free will is real" if one keeps the biological and psychological levels of analysis appropriately separate). After all, humans can commit suicide, if they so desire, which would seem to be the ultimate expression of individual agency. On the other hand, individual human beings and their rational agencies are not only biologically constrained, as are other animals, but also constrained by culturally normative values and reasons about what one ought to do. Suicide seems like a viable option only to individuals who grow up in certain cultures with certain values. So we must somehow recognize in our account of human normative agency both the liberating and the constraining role of coming to maturity in the midst of other persons with whom we are vitally interdependent, both cooperatively and culturally. For the individual, there is no scientific resolution to this essential tension, but the recognition of it constitutes both the source and the energy of much of humans' most profound art and literature—and more than a few psychiatric disorders.

7 Agency as Behavioral Organization

> Our anthropocentric way of looking at things must retreat further and further, and the standpoint of the animal must be the only decisive one.
> —Jakob von Uexküll, *Umwelt und Innenwelt der Tiere*

I have focused here not on the things that organisms do, but rather on how they do them. The issue is not complexity—Nature produces many complex behaviors through her own hardwiring—but control. I am thus especially concerned with things that animals do with a certain amount of behavioral flexibility (whether or not this involves learning). My claim is that organisms are able to flexibly direct and control their actions if, and only if, their underlying psychology is organized agentively, in the manner of a feedback control system.

I thus want to complement the study of organisms' particular cognitive specializations and learning skills (what is typically referred to as animal cognition) with a study of how those specializations and learning skills are organized for behavioral decision-making and control (what we may refer to as animal agency). Virtually all the phenomena of animal agency considered here, including decision-making, planning, inhibition, metacognition, and others, have been studied by comparative psychologists. But their approach has been to consider these as separate cognitive or behavioral skills or domains, each with its own separate mechanism and function. What I have tried to do here is to consider the individual organism as a whole, acting as an agent to make decisions about how to deploy its particular behavioral and cognitive skills in particular contexts. I have thus attempted to provide an account of the overall organization of agentive decision-making and behavioral control—involving integrated tiers of feedback control mechanisms—at several key points along the evolutionary pathway to humans. The result is my vision for an appropriately broad and encompassing evolutionary psychology.

To round out the argument, in this final chapter, I would like to make and discuss six claims about the agentive organization of behavior across species that will serve both to summarize my theoretical model of this major dimension of psychological functioning and to highlight important issues that need further empirical and theoretical investigation. Hopefully, the model as explicated in the main text and in these summary claims provides enough detail that it can be applied to agentive psychological organization across the full range of animal species, including humans.

1. The "backbone" of behavioral agency is feedback control organization

Just as only a few basic Bauplans for animals' bodies have survived the test of evolutionary time, so only a few basic Bauplans for behavioral organization have survived the test of evolutionary time. From the Cambrian explosion some 500 million years ago, most animal bodies have been organized in a bilateral symmetry, and in the case of vertebrate species, there is in addition the central organizing structure of a backbone. The analogy I am proposing here is that the central organizing structure of vertebrate behavior that gives it its characteristic behavioral flexibility—its backbone, so to speak—is feedback control organization. Feedback control organization empowers the individual agent to behave flexibly as needed to solve problems by directing and controlling its actions, and in some cases even planning its actions toward current or future goals and self-regulating behavioral execution from one or another executive tier of operation.

It is telling that when humans attempt to build a machine that acts autonomously, intelligently, and flexibly in the face of unpredictable ecological challenges, pretty much the only organizational architecture used is feedback control organization. This was the point of our exercise in the prospective engineering of the hypothetical leaf vacuum machine. Today, basically all models in artificial life and robotics, as well as those in computational modeling and the philosophy of action and agency, have this same basic architecture: the agent has goals or values, perceptually attends to situations relevant to those goals or values, makes behavioral decisions (and so acts) in light of those goals or values and relevant situations, and observes its actions and their results to make ongoing adjustments as needed. One can for various purposes focus only on some subset of these components, but this overall organization is needed for even the most basic account of agentive action.

Linearly structured stimulus-response organization cannot produce flexible, agentive action. The original prototype for behaviorists' proposed stimulus-response organization was Pavlov's reflex. But, as John Dewey argued in his famous 1896 paper on the reflex arc concept in psychology, reflexes constitute only a small portion of the behavior of vertebrate animals, and when used as a model for flexible, intelligent action, they highlight only two of the elements, perception and action, in a larger structural organization that also includes the goals that the organism is pursuing, as well as perceptual feedback and online behavioral adjustments. No one today takes the behaviorist theoretical paradigm seriously, but theoretical vocabulary and framing are important, and characterizing animal behavior in terms of stimuli and responses, as is still common in the field, tends to efface its deeper organizational principles.

2. The ecological challenges leading to the evolution of behavioral agency all involve one or another form of unpredictability in the environment When the environment is predictable and the evolutionary function is critical, Nature tends to favor hardwiring. In human behavior, we may think of abilities such as breathing and swallowing and other reflexes, which may involve some agentive control in some special cases but under normal circumstances do not rely on complex agentive decision-making. Other species have many more behaviors structured in this hardwired way. But when Nature cannot predict important future contingencies in the environment, so to speak, her solution is to equip the individual to pursue certain goals flexibly by assessing the immediate situations and then choosing the best thing to do (see Veissière et al., 2019).

From the individual agent's point of view, this often means making a decision in the face of one or another type of uncertainty (or, as a special case, one or another type of risk). In principle, many different things may cause an agent to experience uncertainty in its ecological niche, including many different aspects of the physical environment. But, according to my hypothesis, the most important cause of decision-making uncertainties for agentive organisms is other creatures. More specifically:

- For reptiles, most of the uncertainties arise from the behavior of insect prey, successful pursuit of which requires flexible decision-making (which is also sometimes required for successful escape from predators).

- For mammals, most of the new uncertainties arise from the behavior of group mate conspecifics who compete with them for food, creating pressure to make "better" (more efficient) decisions.

- For great apes, most of the new uncertainties arise again from the behavior of group mate conspecifics, but because of their common preference for clumped and difficult-to-access resources, they compete in especially intense ways, creating pressure both to predict the behavior of competitors more accurately and to correct poor decisions before behavioral execution.

- For humans, most of the new uncertainties arise from the challenging behavior of collaborative partners or groups as they attempt to coordinate with them to obtain resources or carry out other complex activities, requiring a whole host of new social-cognitive skills and motivations, as well as new forms of social decision-making and self-regulation.

These uncertainties represent not specific ecological challenges but general types or patterns of ecological challenge, and this requires not just a specific behavioral adaptation but a new type of psychological organization. In other words, certain *types* of ecological challenges create certain *types* of uncertainties in the decision-making individual, which lead to certain *types* of agentive behavioral organization (opening up certain *types* of experiences). And my specific claim is that all these types flow, at least within the vertebrate clade, initially and mainly from how organisms interact with other creatures. Thus an organism's agentive behavioral organization depends on whether it is mainly solitary, mainly competitive with conspecific group mates (via either scramble or contest competition), or mainly collaborative with conspecific group mates.

3. Despite the plethora of specific behavioral and psychological adaptations across species, only a few basic types of psychological Bauplans exist for the agentive organization of behavior   Of course, the specific number of Bauplans we are talking about depends on whether one lumps or splits (creating larger, more general categories or smaller, more detailed ones), with the possibility that the typology I have proposed here can be broken down into a number of more specific types. And I have not systematically considered animal taxa off the pathway to humans. But given this pathway and the differentiation of the four sets of socioecological challenges just outlined, the main types of agents and their key operational features are the following.

- As goal-directed agents who forage for unpredictably behaving insects, reptiles pursue goals relatively flexibly via a series of go-no-go decisions. They can also, when necessary, employ a kind of reactive global inhibition in which they stop what they are doing (e.g., eating) and do something else (e.g., fleeing a predator), thus flexibly changing a "go" to a "no-go" decision, and vice versa, as required by the situation.

- As intentional agents who compete with conspecifics for access to food, mammals operate in the same basic way as reptiles, but they have, in addition, evolved a new executive tier of behavioral organization. This new executive tier enables individuals to intend action plans to goals before enacting them. That is, it enables individuals to pursue goals more efficiently by first engaging in proactive planning and decision-making using cognitive simulations involving a choice between simultaneously available action possibilities, which outputs not an action but an intention to act. Such either-or decision-making leading to intended actions enables a more flexible type of inhibitory control that boosts the value of the chosen action and diminishes the value of the unchosen alternatives.

- As rational agents who engage in especially intense scramble and contest competition with group mates, great apes evolved a new second-order executive tier of behavioral organization that monitors the other tiers. This reflective tier enables individuals to plan even for goals that they are not currently entertaining; to correct decisions that they judge as faulty (e.g., by gathering further information); and to identify and resolve various types of conflicts among goals before acting. This planning and decision-making are logically structured by inferences about intentional and causal relations in the environment, attributed on the basis of the agent's reflective access to its own first-order action planning and decision-making (i.e., it simulates external events based on its own "logic of action"). Great apes thus make their decisions rationally, in at least one definition of the term.

- As normative agents who must collaborate with others to forage successfully, humans evolved the skills and motivations necessary to form with those others either a joint or a collective agency. This enabled them to pursue new goals not accessible to individuals, and it led to the emergence of a host of new cognitive and motivational processes enabling individuals to coordinate mental states with others flexibly and to collaboratively or collectively self-regulate the process by invoking normative standards that were effective in controlling participants' actions, at least partly because those partners see them as objective and so dispositive. Shared agencies create three simultaneously active modes— individual agency, shared agency (either joint or collective), and role agency—which must somehow be harmonized to determine the specific action to be executed.

And so the fourfold typology of behavioral agency proposed here is generated by augmenting the backbone of feedback control organization with either one or two executive tiers of decision-making and control, which may then be reorganized in the case of humans into some form of shared agency as multiple rational agents pool their efforts to pursue novel goals (see table 7.1). Researchers in animal cognition have investigated a number of executive skills, including action planning, inhibitory control, delay of gratification, behavioral updating, and others. But some models in modern cognitive science view these not as separate skills but as interrelated components of one or more overarching tiers of executive monitoring and control (e.g., see Koechlin & Summerfield, 2007). I have therefore adopted the architecture of integrated tiers of operation to emphasize that the various executive processes operate together, in concert, as a control system. The first-order executive tier's goal is to help make "better" decisions, which it does by attending to the operational (behavioral-perceptual) tier of operation, including the agent's goals, actions, and action results. The second-order executive tier, the reflective tier, is likewise a control system, in this case aimed at making *even better* better decisions, which it does by attending to the first-order decision-making process, including discrepancies in goals and in its own information about a situation as it unfolds over time.

Given how much the human species differs from others, it is perhaps surprising that the new machinery involved is somewhat modest. But that is the miracle of it: a seemingly modest change has led to new forms of agency capable of all kinds of new accomplishments. One of the keys is

Table 7.1

Typology of different types of agency and their key characteristics

	Architecture	Directing of Actions [and attention]	Controlling of Actions
Goal-Directed Agents	Feedback Control System	Goals [attention to relevant situations]	Global Inhibition of Go-No-Go Decisions
Intentional Agent	Additional Executive Tier [working memory]	Intentions as Plans to Goals [attention to goals and actions]	Monitoring Uncertainty in Either-Or Decisions
Rational Agent	Additional 2nd Order Executive Tier [metacognition]	Plans for Future Goals [attention to executive decision]	Diagnosing Problems and Intervening to Make "Better" Decisions
Normative Agents	Social Self-Regulation of Executive Tiers [reputation; obligation]	Shared Goals with Meshing Plans [perspectival flexibility]	Social-Normative Self-Regulation [via rational/moral norms]

humans' extremely protracted ontogeny, during which individuals construct many of their cognitive skills as they are collaborating and communicating and self-regulating their shared agencies with others. Coordinating with other individuals toward shared goals requires taking others' role and perspective on things, leading to perspectival and ultimately to objective cognitive representations, accompanied by recursively structured inferences. And interacting with others in all kinds of shared agencies requires individuals to respect one another as equal participants and to trust one another in the "agreements" that they make, from joint commitments to the social norms of the group, leading ultimately to relationships structured by normative attitudes and emotions. As children learn to participate with others in shared agencies, they learn to coordinate simultaneously the three modes of agency potentially involved. The outcome is an individual who has both empowered itself with shared agencies and, at the same time, constrained its own individual agency to coordinate with the numerous shared agencies in which it participates.

These different types of agency emerged sequentially in evolution, from goal-directed to intentional to rational to socially normative, with each building on its predecessor (given that the species on which I have focused form a continuous evolutionary line; see Bonner, 1988, for an account of how more complex biological forms emerge from simpler ones over evolutionary time). And so we might evoke once more our onion metaphor. Each of the derived forms of agency encompasses the earlier forms as inner layers, with new components added and integrated (perhaps transforming things in the process). In the case of humans, for example, some of the simple, sensorimotor things they do are still organized by a goal-directed agency, and some of the more complex things they do involve all the different layers operating at different levels simultaneously.

One might question whether other types of agency are not included in the current typology. Of course, small variations on these themes certainly exist, and it is very likely that some insects and birds—especially the more social species—do things in some somewhat different ways. But given the delimited set of theoretical tools I have used—feedback control organization, executive tiers of functioning, different forms of planning and decision-making, cognitive control—there are not so many other possibilities. However, it could easily be that these component processes might be reconfigured in other ways in other species, and it is also possible that there could be other major types of agency if organisms use components other than those outlined here to direct and control their actions.

4. The evolutionary emergence of new forms of behavioral organization involves both "hierarchical modularity" and "trickle-down selection"
Behaviorism and evolutionary psychology share very little in common. But one commonality is that they both focus on a fairly molecular level of analysis. This is obvious in the case of behaviorism's stimulus-response analysis, which is extremely difficult to apply if one thinks of a normal, everyday animal behavior such as a lizard foraging for insects: how many stimuli and responses are involved in this behavior, and how are they connected to one another? For its part, evolutionary psychology focuses on highly specific computational mechanisms; for example, applied to lizards, it might posit a specific mechanism for computing the flight direction of a fleeing ant. But the relation to the other mechanisms required for prey capture is left unspecified. A flight-computing mechanism can only be fully understood—and its evolution explained—by specifying the nature of its

participation in the larger activity of foraging for insects. A mechanism for computing the flight direction of a fleeing ant would evolve, obviously, not for its own sake but rather to facilitate the hierarchically governing goal of capturing the prey. Keeping this obvious fact in mind is important, because it is necessary for explaining why the behavior of a species evolves in the way that it does—or so I am about to claim.

Let us imagine a situation in which ecological conditions change, and the individuals of a species can no longer procure their normal food. The first step might be natural selection for individuals who already have at least some motivation for another food that is more readily available. This selective process will have myriad "trickle-down" effects on other behavioral components, as the ability to do what it takes to obtain the new food also comes under selection pressures as means to that higher-level end. Thus, if an organism must now climb trees to obtain the newly necessary food, and it previously climbed trees only to escape predators, at least some individuals might figure out how to co-opt their skills of climbing for this new end. This process is most typically called preadaptation (or exaptation), and we can also imagine a similar process in which the organism does not already know how to climb but stretches its existing walking capacities to somehow get up the tree. In either case, the point is that the need to somehow successfully climb trees emanates from the higher-level goal of obtaining the new food, and climbing came into existence with no new genetic events. Then, once this new foraging pattern is present in a number of individuals, subsequent generations of individuals will be selected for genetically based skills that facilitate the new foraging pattern, including climbing (in a process that is most often called "genetic assimilation"). If we think hierarchically, we can often see behavior leading the way in an evolutionary process, as new goals create trickle-down pressures on an individual's agentive powers to adjust, which may then set the stage for a future process of genetic assimilation.

The resulting model of modularity in this approach is what I have called hierarchical modularity (as characteristic of most complex machines). Computing the flight direction of a fleeing insect and climbing a tree may operate just as evolutionary psychology envisions them: as dedicated computational mechanisms. However, these computational mechanisms mostly evolved for higher-level goals, and they may then be co-opted by other higher-level goals, or employed in modified fashion, agentively, for still other higher-level goals. Modularity in this sense is thus about behavioral competencies that are components in the hierarchically organized pursuit of potentially many higher-level goals, with the evolutionary changes necessary for achieving any given goal (including the creation of new modules) "directed" from above. Hierarchical modularity is thus responsible for much of the flexibility in an organism's behavior, as Nature hardwires goals into individuals to make sure certain things get done, but then leaves it up to the individual to figure out how to use its existing competencies to attain those goals. The division of the world into domain-specific and domain-general functions—as the issue is often framed in contemporary debates— does not capture any of these structures or processes very well.

I thus propose that we recognize the hierarchical structure of behavior for all complex organisms, and that our analysis of particular adaptations or behavioral functions always keeps this structure in mind. Whereas for some simple reflexes, hierarchical structure is very likely not applicable, in the vast majority of cases of the type we are concerned with here, hierarchical structure is necessary to explain the evolution of the behavioral flexibility and underlying agentive organization involved.

5. Changes in the agentive organization of a species' behavior and psychology lead to changes in the types of experience it is capable of having (its experiential niche) There is not just one environment for all organisms. We humans may conceptualize one "objective" environment for all organisms (because that is the way we experience the world), but the fact of the matter is that different organisms live in different ecological and experiential niches depending on what they need to do to survive and thrive. As part of the process, the organism perceives, attends to, and knows about situations relevant to these actions. Each organism lives in its own ecological and experiential niche, as determined by its behavioral capabilities.

This fact is obvious when it comes to the specific adaptations of particular species: a bird visually perceives flying insects and swoops in to capture them; a bat echolocates flying insects and uses that to direct its flight to the prey; worms "feel" or smell their food. Most mammals do not distinguish red and green in their visual perception, whereas most primates, who need to distinguish ripe from unripe fruits, distinguish those colors. My proposal here is that something analogous is at work when we consider different kinds of agents: different types of agentive behavioral organization lead to different types of experiential niches. More specifically:

- Reptiles and other goal-directed agents direct their actions toward goals. In doing so, they do not just perceive everything that their perceptual organs take in, but rather actively attend (top down) to situations that are relevant to their goals. Because a goal is a cognitive representation of a desired situation, the organism's attention must be directed not to objects or actions but to environmental situations, specifically those that represent opportunities or obstacles for goal attainment.

- Squirrels and other intentional agents operate not only with an operational (perception-action) tier of functioning but also with an executive tier of decision-making and control. This manner of functioning leads to an experiential niche that includes not just relevant situations in the world but also the individual's own operational level of functioning in terms of its goals, actions, and results. I have thus posited that we may think of mammals and other intentional agents as conscious beings, who not only act toward goals flexibly but, in some sense, know what they are doing.

- Great apes and other rational agents live in an experiential niche in which relevant situations are determined quite often by the causal and intentional relations among entities in the external environment. These relations come into being for apes as they attribute the causality of their own actions to external events and the intentionality of their own thinking and planning to external agents. This attribution process is made possible by a second-order executive (reflective) tier of operation, which gives apes access to their own first-order executive processes. The new experiential niche of great apes thus includes both causal and intentional relations in the external world and their own psychological functioning on the first-order executive tier of planning, decision-making, and cognitive control.

- Human beings operate as normative agents in shared agencies of various types to pursue shared goals, requiring them to mentally coordinate with other rational agents. Operating in this manner over several years of ontogeny leads individuals to construct, as a new and species-unique experiential niche, an objective-normative world—which contrasts with individual subjective perspectives and values—that adjudicates for everyone alike, impartially, what is the correct thing to believe or to do. This way of operating also leads to the objectification of various kinds of "agreements" in the social and institutional structures of a culture (e.g., money, husband), which thereby gain certain deontic powers, as well as to a new form of socially perspectivized self-consciousness.

For me, an understanding of the process of behavioral evolution and how it structures the organism's experiential niche undermines any and all philosophical positions that give a privileged role to the human experiential niche as an accurate, objective description of "the" environment to which all organisms are adapting. We humans can say that there is an objective world, and different organisms perceive different aspects of it, but that just means that we can compare our world to what we believe is their world (perhaps based on scientific evidence). Each species of organism lives in its own ecological niche, and thus in its own experiential niche, and this obviously applies to humans as well, including scientists and philosophers. In any case, be that as it may, the key point is simply that Nature selects most directly for adaptive actions, and this drives everything in the evolution of an organism's psychology, including its experiential niche.

6. The decision-making agent is necessary, and it is not a homunculus, at least not in a bad way Reductionism is an imperative for many scientists and philosophers, and it has its uses, especially in making connections between different fields of study. But sometimes the reduction eliminates the phenomenon of interest. Thus models of cognitive science built on the computer metaphor or neuroscientific explanations often do not have room for a decision-making agent. If a model has a decision-maker that is not somehow mechanical, the charge of "homunculus" is levied, and there is an attempt to eliminate it. But I believe that for the most important phenomena of psychology, the decision-making agent cannot be eliminated without losing those phenomena. An analogy to the history of biology is instructive.

In the eighteenth and nineteenth centuries, as scientists were breaking down all kinds of objects and substances into their chemical and even atomic elements, the question was whether living things were composed of the same elements as everything else. The fact that nonliving things were inert unless acted on, whereas living things spontaneously produced their own actions, suggested to some scientists an élan vital, a kind of vital substance or energy that animated living things. That proved not to be the case. The explanation, as we now know, is that living things are made up of the same substances as nonliving things, but these substances are organized in unique ways so as to transform the energy of, ultimately, the sun in support of the energetic processes of life such as growth, reproduction, and, in the case of animals, behavior.

The form of that explanation can be applied on the psychological level as well. Is an agent some kind of homunculus or élan vital behind the individual's production of actions? No, what lies behind the individual's production of agentive actions is not a new entity such as a homunculus but a particular kind of psychological organization in which a living individual attends to goal-relevant situations, makes decisions, and self-regulates the process. Not all action is produced in this way. We may not need agency to explain the actions of Venus flytraps or bacteria, only the same energetic principles that explain their growth, reproduction, and other life-sustaining functions. But to explain how organisms behave flexibly and efficiently in novel environmental circumstances, we need a new behavioral principle, and that new behavioral principle is an underlying psychological organization of agency based on principles of feedback control. We do not need a mysterious substance, but only the living individual acting on the world through the structure of a particular type of behavioral and psychological organization.

For some scientists and philosophers, it is difficult to conceptualize a psychological decision-maker that is not wholly reducible to biological or physical causality. But this is why I started my account with machines. It is one of the great discoveries of the twentieth century that machines made exclusively of nonliving components can produce actions that are, in many important respects, agent-like. They do this not via any novel components but via a novel form of feedback control organization. This raises the difficult question of whether machines are, or could be, actual agents. If we look only at existing machines, it is reasonable to be skeptical. But then the challenge is to specify what machines might be missing as compared with living organisms, and we are back to homunculi or some such. But I repeat: the analogy to vitalism is instructive. If we ask whether mixing various carbon-related chemicals in a bowl in particularly organized ways could produce life, we are in the same quandary. My own view is that what most clearly differentiates agentive organisms from behaving machines, as they are currently configured, is the way that living agents flexibly attend to *relevant* situations—opportunities and obstacles to their goals and values—that are both structured by, and serve to structure, their goal-directed actions; their perception and cognition are integrally fused with their goal-directed actions. We may look forward to a future, perhaps, in which machines organized in this more integrated manner are used to facilitate human activities of all kinds.

I thus believe that deploying psychological mechanisms effectively in novel situations—for which evolutionary processes cannot prepare the individual in detail—requires an agent. Indeed, that is the whole point. Because it cannot predict the particularities of the future situations that an individual might encounter, Nature has constructed an underlying psychological organization of agency enabling the individual to make its own decisions and self-regulate its own actions in pursuit of goals that, ultimately, Nature has built in. The concept of agency thus involves both deterministic and spontaneous elements. An organism's capacities for acting agentively come into existence phylogenetically through processes of evolution by means of natural selection and ontogenetically through processes of (epi)genetic expression; they are thereby, in an important sense, determined. But these capacities must still be exercised in the moment, and for that we need a psychological agent whose defining feature is the making of spontaneous and independent behavioral choices. Nature may determine my capacities for using a language, but it does not determine what I say. Through normal processes of evolution by means of natural selection, Nature has crafted forms of agentive organization that empower individuals to act autonomously.

So my final proposal is this. Every scientific discipline begins with a proper domain, a first principle, as Aristotle would call it. In biology, that proper domain or first principle is life: physical substances organized in particular ways to perform particular organismic functions. In psychology, depending on one's theoretical predilections, that proper domain or first principle might be either behavior or mentality. But my preferred candidate would be agency, precisely because agency is the organizational framework within which both behavioral and mental processes operate. In the theoretical model I am proposing, agency can take a specifiable range of forms, which vary along a specifiable number of dimensions, and this provides a common set of analytic tools for constructing more specific models of behavioral organization across animal species. Or at least that is the program.

Addendum A

Despite its demise as a theory to which anyone explicitly subscribes, behaviorism has influenced the linguistic framing with which many researchers still talk about animal behavior. Basically, the language of stimulus and response adopts the experimenter's point of view on the experimental situation, given that she knows the organism's goal (in the experimental situation, usually obtaining food). She presents the organism with a stimulus (a situation discrepant with its goal, e.g., out-of-reach food), with the organism's perception or understanding of the stimulus not an issue. Control theory begins with the organism's goal(s) and its perception of, and attention to, the situation. It understands the organism not as responding to a stimulus but as seeking to achieve goals and using its perception and attention to the situation to make effective behavioral decisions (and learning from feedback which actions are best at achieving which goals in which situations). The following table summarizes the alternative terminologies.

Behaviorist terminology	Control theory terminology
Stimulus	Perception of, or attention to, the situation from the organism's point of view. If we desire the experimenter's point of view, we may say things like "the presented object or situation."
Response	Action or behavior that the organism chooses as a means for realizing its goal.
Reinforcement or reward	Goal attainment, which is reinforcing because the perceived situation now matches the internal goal.

Behaviorist terminology	Control theory terminology
(a) Learning a response to a stimulus (b) Learning an association between stimuli	(a) Learning the actions necessary to attain a goal in a particular situation (from feedback of results of actions). (b) Learning the various relations among perceived entities and events. Decision-making Executive processes: (i) reactive (e.g., inhibition); (ii) proactive (e.g., planning); (iii) behavior monitoring and control (feedback) Second-order executive processes (reflection, metacognition)

Addendum B

My choices for model species were to some degree influenced by the existence of relevant experimental data. Nevertheless, in some cases, the data were still thinner than would have been ideal. The problem is not that there are not enough data on animal behavior in general, but that the existing data are often not of the right kind to answer the questions I am asking. Thus the many data from behavioral ecology and associative learning theory are interesting and important in their own right, but they often do not address issues of direct relevance to cognition and agency. It is thus important for researchers in animal cognition—with their more psychologically attuned methods and perspectives—to expand as widely as possible the species they investigate. But any one researcher can do only so much, given the obvious limitations of time, resources, and access to relevant species. As one approach to these problems, an interesting initiative—although at the moment restricted to primates—is a project in which different animal researchers can propose comparative experiments on a website, and then other researchers can collect data with their various focal species and contribute (Many Primates, https://manyprimates.github.io; see also Atlas of Comparative Cognition, https://acc.clld.org). In this way, animal researchers may pool their expertise and resources to conduct larger-scale comparative studies focused on particular psychological processes (e.g., as in E. MacLean et al., 2014, chap. 3, which provides a kind of proof of concept for the project going beyond primates; see also Many Primates, 2019a, 2019b).

Notes

Chapter 1

1. The most strident criticisms of Wilson's attempt were political. Given the history of racism and eugenics in America and Germany in the first half of the twentieth century, the proposal that human social behavior has genetic bases was politically explosive.

Chapter 2

1. Technically, the action, like a furnace's action, occurs outside the control system; it is what is being controlled. So the third component is the process of decision-making about what action to perform, what I am calling here the behavioral decision.

Chapter 3

1. This is all so-called top-down attention, driven by the organism's internally represented goals and values. The other form is typically called bottom-up attention, which is more stimulus driven, for example, a loud noise that attracts attention involuntarily. But one could also think of bottom-up attention as driven by Nature's goals or reference values. Thus we involuntarily attend to a sudden loud noise because Nature selects us to quickly see if the noise is coming from a source that has relevance for basic biological "goals" like survival. So even involuntary bottom-up attention could be said to be, from the perspective of Nature, also goal relevant.

2. Perhaps the structuring of experience in terms of relevant situations is a kind of precursor to the human conceptualization of experience in terms of propositions, and thus what Wittgenstein was referencing when he opened his 1921 book by stating: "The world is all that is the case. The world is the totality of facts, not of things."

Chapter 4

1. For some nonscientific but nevertheless informative (and entertaining) observations of squirrels' behavioral flexibility, check out any of the many home videos on YouTube capturing the ways that backyard squirrels can outsmart human homeowners and their bird feeders.

2. In Michotte's (1963) classic studies, humans perceived a causal link between external events whenever there was the exact same kind of spatial-temporal contiguity (a second or so) that prompts rats to perceive their own acts as causing results in these studies.

References

Allritz, M., Call, J., & Borkenau, P. (2015). How chimpanzees perform in a modified emotional Stroop task. *Animal Cognition, 19*, 1–15.

Amici, F., Aureli, F., & Call, J. (2008). Fission-fusion dynamics, behavioral flexibility, and inhibitory control in primates. *Current Biology, 18*(18), 1415–1419.

Aristotle. (1984). Politics. In J. Barnes (Ed.), *The complete works of Aristotle*. Princeton, NJ: Princeton University Press.

Aron, A., Robbins, T., & Poldrack, R. (2014). Inhibition and the right inferior frontal cortex: One decade on. *Trends in Cognitive Science, 18*, 177–185.

Ashby, W. R. (1952). *Design for a brain*. London: Chapman and Hall.

Baker, C. L., Saxe, R., & Tenenbaum, J. B. (2009). Action understanding as inverse planning. *Cognition, 113*(3), 329–349.

Banich, M. T. (2009). Executive function: The search for an integrated account. *Current Directions in Psychological Science, 18*(2), 89–94.

Basile, B. M., Schroeder, G. R., Brown, E. K., Templer, V. L., & Hampton, R. R. (2015). Evaluation of seven hypotheses for metamemory performance in rhesus monkeys. *Journal of Experimental Psychology: General, 144*(1), 85.

Baumard, N., André, J. B., & Sperber, D. (2013). A mutualistic approach to morality: The evolution of fairness by partner choice. *Behavioral and Brain Sciences, 36*(1), 59–78.

Bechtel, W., & Bich, L. (2021). Grounding cognition: Heterarchical control mechanisms in biology. *Philosophical Transactions of the Royal Society B: Biological Sciences, 376*, 20190751.

Begun, D. R. (2003). Planet of the apes. *Scientific American, 289*(2), 74–83.

Berkman, E. T., Hutcherson, C. A., Livingston, J. L., Kahn, L. E., & Inzlicht, M. (2017). Self-control as value-based choice. *Current Directions in Psychological Science, 26*, 422–428.

Bermudez, J. (2003). *Thinking without words*. Oxford: Oxford University Press.

Blaser, R., & Ginchansky, R. (2012). Route selection by rats and humans in a navigational traveling salesman problem. *Animal Cognition, 15*, 239–250.

Boehm, C. (1999). *Hierarchy in the forest: The evolution of egalitarian behavior*. Cambridge, MA: Harvard University Press.

Boesch, C. (1994). Cooperative hunting in wild chimpanzees. *Animal Behaviour, 48*(3), 653–667.

Bohn, M., Allritz, M., Call, J., & Völter, C. J. (2017). Information seeking about tool properties in great apes. *Scientific Reports, 7*(1), 1–6.

Bohn, M., Call, J., & Tomasello, M. (2016). The role of past interactions in great apes' communication about absent entities. *Journal of Comparative Psychology, 130*, 351–357.

Bonner, J. T. (1988). *The evolution of complexity by means of natural selection*. Princeton, NJ: Princeton University Press.

Boucherie, P. H., Loretto, M. C., Massen, J. J., & Bugnyar, T. (2019). What constitutes "social complexity" and "social intelligence" in birds? Lessons from ravens. *Behavioral Ecology and Sociobiology, 73*(1), 12.

Bowles, S., & Gintis, H. (2011). *A cooperative species: Human reciprocity and its evolution*. Princeton, NJ: Princeton University Press.

Boyd, R., & Richerson, P. J. (1985). *Culture and the evolutionary process*. Chicago: University of Chicago Press.

Boyd, R., & Richerson, P. J. (2005). *The origin and evolution of cultures*. Oxford: Oxford University Press.

Bradley, B. (2020). *Darwin's psychology*. Oxford: Oxford University Press.

Bratman, M. (1987). *Intention, plans, and practical reason*. Cambridge, MA: Harvard University Press.

Bratman, M. (2014). *Shared agency: A planning theory of acting together*. New York: Oxford University Press.

Braver, T. S. (2012). The variable nature of cognitive control: A dual mechanisms framework. *Trends in Cognitive Sciences, 16*(2), 106–113.

Bray, E. E., MacLean, E. L., & Hare, B. A. (2014). Context specificity of inhibitory control in dogs. *Animal Cognition, 17*(1), 15–31.

Brown, R., Lau, H., & LeDoux, J. (2019). Understanding the higher-order approach to consciousness. *Trends in Cognitive Science, 23*, 754–768.

Bruner, J. S. (1973). Organization of early skilled action. *Child Development, 44*, 1–11.

Burghardt, G. M. (1966). Stimulus control of the prey attack response in naive garter snakes. *Psychonomic Science, 4*(1), 37–38.

Buttelmann, D., Carpenter, M., Call, J., & Tomasello, M. (2007). Enculturated apes imitate rationally. *Developmental Science, 10*, 31–38.

Buttelmann, D., Carpenter, M., Call, J., & Tomasello, M. (2008). Rational tool use and tool choice in human infants and great apes. *Child Development, 79*, 609–626.

Call, J. (2004). Inferences about the location of food in the great apes (*Pan paniscus, Pan troglodytes, Gorilla gorilla,* and *Pongo pygmaeus*). *Journal of Comparative Psychology, 118*(2), 232–241.

Call, J. (2010). Do apes know that they could be wrong? *Animal Cognition, 13*(5), 689–700.

Call, J., & Carpenter, M. (2001). Do apes and children know what they have seen? *Animal Cognition, 3*(4), 207–220.

Call, J., Hare, B., Carpenter, M., & Tomasello, M. (2004). "Unwilling" versus "unable": Chimpanzees' understanding of human intentional action. *Developmental Science, 7*(4), 488–498.

Call, J., & Tomasello, M. (2007). The gestural repertoire of chimpanzees (*Pan troglodytes*). In J. Call & M. Tomasello (Eds.), *The gestural communication of apes and monkeys* (pp. 17–39). Mahwah, NJ: Lawrence Erlbaum Associates.

Carpenter, M., Nagell, K., & Tomasello, M. (1998). Social cognition, joint attention, and communicative competence from 9 to 15 months of age. *Monographs of the Society for Research in Child Development, 63*(4), i–174.

Cheney, D. L., & Seyfarth, R. M. (1991). Reading minds or reading behaviour? Tests for a theory of mind in monkeys. In A. Whiten (Ed.), *Natural theories of mind: Evolution, development and simulation of everyday mindreading* (pp. 175–194). Cambridge: Blackwell.

Chow, P., Leaver, L., Wang, M., & Lea, S. (2015). Serial reversal learning in grey squirrels: Learning efficiency as a function of learning and change of tactics. *Journal of Experimental Psychology: Animal Learning and Cognition, 41*, 343–353.

Chow, P. K. Y., Lea, S. E., de Ibarra, N. H., & Robert, T. (2019). Inhibitory control and memory in the search process for a modified problem in grey squirrels, *Sciurus carolinensis. Animal Cognition, 22*(5), 645–655.

Clark, A. (2015). *Surfing uncertainty. Prediction, action, and the embodied mind.* Oxford: Oxford University Press.

Collingwood, R. G. (1940). *Essay on metaphysics.* London: Oxford University Press.

Cooper, W., Pérez-Mellado, V., & Hawlena, D. (2007). Number, speeds, and approach paths of predators affect escape behavior by the Balearic lizard, *Podarcis lilfordi*. *Journal of Herpetology*, *41*(2), 197–204.

Coqueugniot, H., Hublin, J.-J., Veillon, F., Houet, F., & Jacob, T. (2004). Early brain growth in *Homo erectus* and implications for cognitive ability. *Nature*, *231*, 299–302.

Crystal, J. (2013). Remembering the past and planning for the future in rats. *Behavioural Processes*, *93*, 39–49.

Csibra, G., & Gergely, G. (2009). Natural pedagogy. *Trends in Cognitive Sciences*, *13*(4), 148–153.

Custance, D. M., Whiten, A., & Bard, K. A. (1995). Can young chimpanzees (*Pan troglodytes*) imitate arbitrary actions? Hayes & Hayes (1952) revisited. *Behaviour*, *132*(11–12), 837–859.

Darwin, C. (1859). *On the origin of species*. London: John Murray.

Darwin, C. (1871). *The descent of man and selection in relation to sex*. Facsimile of the first edition, ed. Princeton University Press. Original edition, John Murray, London, 1871.

Davidson, D. (2001). *Subjective, intersubjective, objective*. Oxford: Oxford University Press.

Dawkins, R. (1976). *The selfish gene*. Oxford: Oxford University Press.

Dawkins, R. (1986). *The blind watchmaker*. New York: Norton.

Dewey, J. (1896). The reflex arc concept in psychology. *Psychological Review*, *3*(4), 357–370.

Dewey, J. (1916). *Democracy and education: An introduction to the philosophy of education*. New York: Macmillan.

Diamond, A. (2013). Executive functions. *Annual Review of Psychology*, *64*, 135–168.

Dickinson, A. (2001). Causal learning: An associative analysis (the 28th Bartlett Memorial Lecture). *Quarterly Journal of Experimental Psychology*, *54B*, 3–25.

Donahue, C. J., Glasser, M. F., Preuss, T. M., Rilling, J. K., & Van Essen, D. C. (2018). Quantitative assessment of prefrontal cortex in humans relative to nonhuman primates. *Proceedings of the National Academy of Sciences*, *115*(22), E5183–E5192.

Duguid, S., Wyman, E., Bullinger, A. F., Herfurth, K., & Tomasello, M. (2014). Coordination strategies of chimpanzees and human children in a stag hunt game. *Proceedings of the Royal Society B: Biological Sciences*, *281*(1796), 20141973.

Dunham, Y. (2018). Mere membership. *Trends in Cognitive Sciences*, *22*(9), 780–793.

Egner, T. (2017). *The Wiley handbook of cognitive control*. Hoboken, NJ: Wiley-Blackwell.

Fletcher, G., Warneken, F., & Tomasello, M. (2012). Differences in cognitive processes underlying the collaborative activities of children and chimpanzees. *Cognitive Development, 27*(2), 136–153.

Foote, A. L., & Crystal, J. D. (2007). Metacognition in the rat. *Current Biology, 17*(6), 551–555.

Frijda, N. (1986). *The emotions*. Cambridge: Cambridge University Press.

Geertz, C. (1973). *The interpretation of cultures*. New York: Basic Books.

Gershman, S. J., Horvitz, E. J., & Tenenbaum, J. B. (2015). Computational rationality: A converging paradigm for intelligence in brains, minds, and machines. *Science, 349*, 273–278.

Gibson, J. J. (1977). The concept of affordances. In R. Shaw & J. Bransford (Eds.), *Perceiving, acting, and knowing* (pp. 67–82). Hillsdale, NJ: Lawrence Erlbaum.

Gigerenzer, G., Hertwig, R., & Pachur, T. (2011). *Heuristics: The foundation of adaptive behavior*. Oxford: Oxford University Press.

Gigerenzer, G., & Selten, R. (2001). *Bounded rationality: The adaptive toolbox*. Cambridge, MA: MIT Press.

Gigerenzer, G., & Todd, P. (1999). *Simple heuristics that make us smart*. Oxford: Oxford University Press.

Gilbert, M. (2014). *Joint commitment: How we make the social world*. New York: Oxford University Press.

Gilby, I. C., & Wrangham, R. W. (2007). Risk-prone hunting by chimpanzees (*Pan troglodytes schweinfurthii*) increases during periods of high diet quality. *Behavioral Ecology and Sociobiology, 61*(11), 1771–1779.

Godfrey-Smith, P. (2016). *Other minds: The octopus, the sea, and the deep origins of consciousness*. New York: Farrar, Straus and Giroux.

Godfrey-Smith, P. (2020). *Metazoa: Animal life and the birth of the mind*. New York: Farrar, Straus and Giroux.

González-Forero, M., & Gardner, A. (2018). Inference of ecological and social drivers of human brain-size evolution. *Nature, 557*, 554–557.

Gordon, R. (in press). Simulation, predictive coding, and the shared world. In K. Ochsner & M. Gilead (Eds.), *The neural basis of mentalizing*. Springer.

Gräfenhain, M., Behne, T., Carpenter, M., & Tomasello, M. (2009). Young children's understanding of joint commitments. *Developmental Psychology, 45*, 1430–1443.

Graziano, M. S. (2019). *Rethinking consciousness*. New York: Norton.

Greenberg, J. R., Hamann, K., Warneken, F., & Tomasello, M. (2010). Chimpanzee helping in collaborative and non-collaborative contexts. *Animal Behaviour, 80*(5), 873–880.

Gunz, P., Tilot, A. K., Wittfeld, K., Teumer, A., Shapland, C. Y., Van Erp, T. G., et al. (2019). Neandertal introgression sheds light on modern human endocranial globularity. *Current Biology, 29*(1), 120–127.

Halina, M., Rossano, F., & Tomasello, M. (2013). The ontogenetic ritualization of bonobo gestures. *Animal Cognition, 16*, 653–666.

Hamann, K., Warneken, F., Greenberg, J. R., & Tomasello, M. (2011). Collaboration encourages equal sharing in children but not in chimpanzees. *Nature, 476*(7360), 328–331.

Hamann, K., Warneken, F., & Tomasello, M. (2012). Children's developing commitments to joint goals. *Child Development, 83*, 137–145.

Hanus, D., & Call, J. (2008). Chimpanzees infer the location of a reward on the basis of the effect of its weight. *Current Biology, 18*(9), R370–R372.

Hanus, D., & Call, J. (2011). Chimpanzee problem-solving: Contrasting the use of causal and arbitrary cues. *Animal Cognition, 14*(6), 871–878.

Harari, Y. N. (2015). *What explains the rise of humans?* [Video]. TEDGlobal London.

Hardecker, S., Schmidt, M. F., & Tomasello, M. (2017). Children's developing understanding of the conventionality of rules. *Journal of Cognition and Development, 18*(2), 163–188.

Hare, B., Call, J., Agnetta, B., & Tomasello, M. (2000). Chimpanzees know what conspecifics do and do not see. *Animal Behaviour, 59*, 771–785.

Hare, B., Call, J., & Tomasello, M. (2001). Do chimpanzees know what conspecifics know? *Animal Behaviour, 61*(1), 139–151.

Hare, B., Wobber, V., & Wrangham, R. (2012). The self-domestication hypothesis: Evolution of bonobo psychology is due to selection against aggression. *Animal Behaviour, 83*(3), 573–585.

Hart, A. (2006). Behavior. In Victor Ambros (Ed.), *WormBook*. The C. elegans Research Community.

Haun, D. B., Nawroth, C., & Call, J. (2011). Great apes' risk-taking strategies in a decision making task. *PLOS ONE, 6*(12), e28801.

Haun, D., & Over, H. (2015). Like me: A homophily-based account of human culture. In T. Breyer (Ed.), *Epistemological dimensions of evolutionary psychology* (pp. 117–130). New York: Springer.

Haun, D., & Tomasello, M. (2011). Conformity to peer pressure in preschool children. *Child Development, 82*, 1759–1767.

Haun, D., & Tomasello, M. (2014). Children conform to the behavior of peers; great apes stick with what they know. *Psychological Science, 25*, 2160–2167.

Heilbronner, S. R., Rosati, A. G., Stevens, J. R., Hare, B., & Hauser, M. D. (2008). A fruit in the hand or two in the bush? Divergent risk preferences in chimpanzees and bonobos. *Biology Letters, 4*(3), 246–249.

Heinol, A., & Martindale, M. Q. (2008). Acoel development supports a simple planula-like urbilaterian. *Philosophical Transactions of the Royal Society B: Biological Sciences, 363*(1496), 1493–1501.

Henrich, J. (2016). *The secret of our success: How culture is driving human evolution, domesticating our species, and making us smarter.* Princeton, NJ: Princeton University Press.

Herrmann, E., Misch, A., Hernandez-Lloreda, V., & Tomasello, M. (2015). Uniquely human self-control begins at school age. *Developmental Science, 18*(6), 979–993.

Herrmann, E., & Tomasello, M. (2015). Focusing and shifting attention in human children (*Homo sapiens*) and chimpanzees (*Pan troglodytes*). *Journal of Comparative Psychology, 129*(3), 268–274.

Hintze, A., Olson, R. S., Adami, C., & Hertwig, R. (2015). Risk sensitivity as an evolutionary adaptation. *Scientific Reports, 5*, 8242.

James, W. (1890). *The principles of psychology* (Vol. 1). New York: Holt.

Jensen, K., Call, J., & Tomasello, M. (2007). Chimpanzees are rational maximizers in an ultimatum game. *Science, 318*, 107–109.

John, M., Duguid, S., Tomasello, M., & Melis, A. P. (2019). How chimpanzees (*Pan troglodytes*) share the spoils with collaborators and bystanders. *PLOS ONE 14*(9), e0222795.

Johnson-Ulrich, L., & Holekamp, K. E. (2020). Group size and social rank predict inhibitory control in spotted hyaenas. *Animal Behaviour, 160*, 157–168.

Juechems, K., & Summerfield, C. (2019). Where does value come from? *Trends in Cognitive Sciences, 23*(10), 836–850.

Juszczak, G. R., & Miller, M. (2016). Detour behavior of mice trained with transparent, semitransparent and opaque barriers. *PLOS ONE, 11*(9), e0162018.

Kaas, J. H. (2013). The evolution of brains from early mammals to humans. *Wiley Interdisciplinary Reviews: Cognitive Science, 4*(1), 33–45.

Kabadayi, C., Bobrowicz, K., & Osvath, M. (2018). The detour paradigm in animal cognition. *Animal Cognition, 21*(1), 21–35.

Kacelnik, A., & El Mouden, C. (2013). Triumphs and trials of the risk paradigm. *Animal Behaviour, 86*(6), 1117–1129.

Kachel, U., Svetlova, M., & Tomasello, M. (2018). Three-year-olds' reactions to a partner's failure to perform her role in a joint commitment. *Child Development, 89,* 1691–1703.

Kachel, U., Svetlova, M., & Tomasello, M. (2019). Three- and 5-year-old children's understanding of how to dissolve a joint commitment. *Journal of Experimental Child Psychology, 184*, 34–47.

Kachel, U., & Tomasello, M. (2019). 3- and 5-year-old children's adherence to explicit and implicit joint commitments. *Developmental Psychology, 55*, 80–88.

Kahneman, D. (2011). *Thinking, fast and slow*. New York: Macmillan.

Kano, F., Krupenye, C., Hirata, S., Tomonaga, M., & Call, J. (2019). Great apes use self-experience to anticipate an agent's action in a false-belief test. *Proceedings of the National Academy of Sciences, 116*(42), 20904–20909.

Karg, K., Schmelz, M., Call, J., & Tomasello, M. (2015a). Chimpanzees strategically manipulate what others can see. *Animal Cognition, 18*, 1069–1076.

Karg, K., Schmelz, M., Call, J., & Tomasello, M. (2015b). The goggles experiment: Can chimpanzees use self-experience to infer what a competitor can see? *Animal Behaviour, 105*, 211–221.

Karmiloff-Smith, A. (1992). *Beyond modularity: A developmental perspective on cognitive science*. Cambridge, MA: MIT Press.

Keijzer, F. (2021). Demarcating cognition: The cognitive life sciences. *Synthese, 198* (Suppl. 1), S137–S157.

Keupp, S., Behne, T., & Rakoczy, H. (2013). Why do children overimitate? Normativity is crucial. *Journal of Experimental Child Psychology, 116*(2), 392–406.

Koechlin, E., & Summerfield, C. (2007). An information theoretical approach to prefrontal executive function. *Trends in Cognitive Sciences, 11*, 229–235.

Lea, S. E., Chow, P. K., Leaver, L. A., & McLaren, I. P. (2020). Behavioral flexibility: A review, a model, and some exploratory tests. *Learning and Behavior, 48*, 1–15.

Leal, M., & Powell, B. J. (2012). Behavioural flexibility and problem-solving in a tropical lizard. *Biology Letters, 8*(1), 28–30.

Lewis, D. (1969). *Convention*. Cambridge, MA: Harvard University Press.

Li, L., Britvan, B., & Tomasello, M. (2021). Young children conform more to norms than to preferences. *PLOS ONE, 16*(5), e0251228.

Liebal, K., Carpenter, M., & Tomasello, M. (2013). Young children's understanding of cultural common ground. *British Journal of Developmental Psychology*, *31*(1), 88–96.

Liebal, K., Pika, S., Call, J., & Tomasello, M. (2004). To move or not to move: How apes adjust to the attentional state of others. *Interaction Studies*, *5*, 199–219.

List, C. (2019). *Why free will is real*. Cambridge, MA: Harvard University Press.

List, C., & Pettit, P. (2011). *Group agency*. Oxford: Oxford University Press.

Lohmann, H., & Tomasello, M. (2003). The role of language in the development of false belief understanding: A training study. *Child Development*, *74*(4), 1130–1144.

Lyon, P., Keijzer, F., Arendt, D., & Levin, M. (2021). Reframing cognition: Getting down to biological basics. *Philosophical Transactions of the Royal Society B: Biological Sciences*, *376*, 20190750.

MacLean, E. L., Hare, B., Nunn, C. L., Addessi, E., Amici, F., Anderson, R. C., et al. (2014). The evolution of self-control. *Proceedings of the National Academy of Sciences*, *111*(20), E2140–E2148.

MacLean, E., Sandel, A., Bray, J., Oldenkamp, R., Reddy, R., & Hare, B. (2013). Group size predicts social but not nonsocial cognition in lemurs. *PLOS ONE*, *8*(6), e66359.

MacLean, P. D. (1990). *The triune brain in evolution: Role in paleocerebral functions*. New York: Springer Science and Business Media.

Mäki-Marttunen, V., Hagen, T., & Espeseth, T. (2019). Proactive and reactive modes of cognitive control can operate independently and simultaneously. *Acta Psychologica*, *199*, 102891.

Manrique, H. M., Gross, A. N. M., & Call, J. (2010). Great apes select tools on the basis of their rigidity. *Journal of Experimental Psychology: Animal Behavior Processes*, *36*(4), 409–422.

Many Primates. (2019a). Collaborative open science as a way to reproducibility and new insights in primate cognition research. *Japanese Psychological Review 62*(3), 205–220.

Many Primates. (2019b). Establishing an infrastructure for collaboration in primate cognition research. *PLOS ONE*, *14*(10), e0223675.

Maynard-Smith, J., & Szathmáry, E. (1995). *The major transitions in evolution*. Oxford: Oxford University Press.

McGrew, W. C. (2010). Chimpanzee technology. *Science*, *328*(5978), 579–580.

McKinney, M. A., Schlesinger, C. A., & Pavey, C. R. (2014). Foraging behaviour of the endangered Australian skink (*Liopholis slateri*). *Australian Journal of Zoology*, *62*(6), 477–482.

Mead, G. H. (1934). *Mind, self, and society*. Chicago: University of Chicago Press.

Melis, A., Altricher, K., Schneider, A., & Tomasello, M. (2013). Allocation of resources to collaborators and free-riders by 3-year-olds. *Journal of Experimental Child Psychology*, *114*, 364–370.

Melis, A. P., Call, J., & Tomasello, M. (2006). Chimpanzees (*Pan troglodytes*) conceal visual and auditory information from others. *Journal of Comparative Psychology*, *120*(2), 154–162.

Melis, A. P., Hare, B., & Tomasello, M. (2006). Engineering cooperation in chimpanzees: Tolerance constraints on cooperation. *Animal Behaviour*, *72*(2), 275–286.

Melis, A. P., Hare, B., & Tomasello, M. (2009). Chimpanzees coordinate in a negotiation game. *Evolution and Human Behavior*, *30*(6), 381–392.

Melis, A., Schneider, A., & Tomasello, M. (2011). Chimpanzees share food in the same way after individual and collaborative food acquisition. *Animal Behaviour*, *82*, 485–493.

Mendelson, T. C., Fitzpatrick, C. L., Hauber, M. E., Pence, C. H., Rodríguez, R. L., Safran, R. J., et al. (2016). Cognitive phenotypes and the evolution of animal decisions. *Trends in Ecology and Evolution*, *31*(11), 850–859.

Mendes, N., Hanus, D., & Call, J. (2007). Raising the level: Orangutans use water as a tool. *Biology Letters*, *3*(5), 453–455.

Michotte, A. (1963). *The perception of causality*. New York: Basic Books.

Miller, G. A., Galanter, E., & Pribram, K. H. (1960). *Plans and the structure of behavior*. New York: Holt.

Möller, R. (2012). A model of ant navigation based on visual prediction. *Journal of Theoretical Biology*, *305*, 118–130.

Molnár, Z. (2011). Evolution of cerebral cortical development. *Brain, Behavior and Evolution*, *78*(1), 94–107.

Mulcahy, N. J., & Call, J. (2006). Apes save tools for future use. *Science*, *312*, 1038–1040.

Nagel, T. (1986). *The view from nowhere*. Oxford: Oxford University Press.

Naumann, R. K., Ondracek, J. M., Reiter, S., Shein-Idelson, M., Tosches, M. A., Yamawaki, T. M., & Laurent, G. (2015). The reptilian brain. *Current Biology*, *25*(8), R317–R321.

Okasha, S. (2018). *Agents and goals in evolution*. Oxford: Oxford University Press.

O'Madagain, C., Schmidt, M., Call, J., Helming, K., Shupe, E., & Tomasello, M. (in press). Apes and children rationally monitor their decisions—but differently. *Proceedings of the Royal Society B*.

Peirce, C. S. (1931). *Collected papers of Charles Sanders Peirce* (C. Hartshorne & P. Weiss, Eds.) (Vol. 1). Cambridge, MA: Harvard University Press.

Piaget, J. (1952). *The origins of intelligence in children.* New York: Norton.

Piaget, J. (1974). *Understanding causality.* New York: Norton.

Piaget, J. (1976). *Le comportement, moteur de l'évolution* (Vol. 354). Paris: Gallimard.

Povinelli, D. J., & Dunphy-Lelii, S. (2001). Do chimpanzees seek explanations? Preliminary comparative investigations. *Canadian Journal of Experimental Psychology, 55*(2), 185.

Powers, W. (1973). *Behavior: The control of perception.* Chicago: Aldine.

Qi, Y., Noble, D. W., Fu, J., & Whiting, M. J. (2018). Testing domain general learning in an Australian lizard. *Animal Cognition, 21*(4), 595–602.

Qin, J., & Wheeler, A. R. (2007). Maze exploration and learning in *C. elegans. Lab on a Chip, 7*(2), 186–192.

Rakoczy, H., Kaufmann, M., & Lohse, K. (2016). Young children understand the normative force of standards of equal resource distribution. *Journal of Experimental Child Psychology, 150*, 396–403.

Redish, A. D. (2016). Vicarious trial and error. *Nature Reviews Neuroscience, 17*(3), 147–159.

Richerson, P. J., & Boyd, R. (2005). *Not by genes alone: How culture transformed human evolution.* Chicago: University of Chicago Press.

Rochat, P. (2021). *Moral acrobatics.* Oxford: Oxford University Press.

Roberts, W., McMillan, N., Musolino, E., & Cole, M. (2012). Information seeking in animals: Metacognition. *Comparative Cognition and Behavior Reviews, 7*, 85–109.

Romain, A., Broihanne, M.-H., De Marco, A., Ngoubangoye, B., Call, J., Rebout, N., & Dufour, V. (2021). Non-human primates use combined rules when deciding under ambiguity. *Philosophical Transactions of the Royal Society B: Biological Sciences.*

Rosati, A. G. (2017a). Foraging cognition: Reviving the ecological intelligence hypothesis. *Trends in Cognitive Sciences, 21*, 691–702.

Rosati, A. G. (2017b). The evolution of primate executive function: From response control to strategic decision-making. In J. Kaas & L. Krubitzer (Eds.), *Evolution of Nervous Systems* (Vol. 3, pp. 423–437). Amsterdam: Elsevier.

Rosati, A. G. (2017c). Decision-making under uncertainty: Preferences, biases, and choice. In *APA handbook of comparative psychology: Perception, learning, and cognition* (Vol. 2, pp. 329–357). American Psychological Association.

Rosati, A. G., & Hare, B. (2011). Chimpanzees and bonobos distinguish between risk and ambiguity. *Biology Letters, 7*(1), 15–18.

Rosati, A. G., & Santos, L. R. (2016). Spontaneous metacognition in rhesus monkeys. *Psychological Science, 27*(9), 1181–1191.

Rosati, A. G., & Stevens, J. R. (2009). The adaptive nature of context-dependent choice. In S. Watanabe, A. Young, L. Huber, A. Blaisdell, & Y. Yamazaki (Eds.), *Rational Animal, Irrational Human* (pp. 101–117). Tokyo: Keio University Press.

Rosati, A. G., Stevens, J. R., Hare, B., & Hauser, M. D. (2007). The evolutionary origins of human patience: Temporal preferences in chimpanzees, bonobos, and human adults. *Current Biology, 17*, 1663–1668.

Santos, L. R., Nissen, A. G., & Ferrugia, J. A. (2006). Rhesus monkeys know what others can and cannot hear. *Animal Behaviour, 71*, 1175–1181.

Santos, L. R., & Rosati, A. G. (2015). The evolutionary roots of human decision-making. *Annual Review of Psychology, 66*, 321–347.

Schelling, T. C. (1960). *The strategy of conflict*. Cambridge, MA: Harvard University Press.

Schmelz, M., Call, J., & Tomasello, M. (2013). Chimpanzees predict that a competitor's preference will match their own. *Biology Letters, 9*(1), 20120829.

Schmidt, M. F. H., Rakoczy, H., & Tomasello, M. (2012). Young children enforce social norms selectively depending on the violator's group affiliation. *Cognition, 124*(3), 325–333.

Schmidt, M., & Tomasello, M. (2012). Young children enforce social norms. *Current Directions in Psychological Science, 21*, 232–236.

Scholz, M., Dinner, A. R., Levine, E., & Biron, D. (2017). Stochastic feeding dynamics arise from the need for information and energy. *Proceedings of the National Academy of Sciences, 114*(35), 9261–9266.

Searle, J. (1995). *The construction of social reality*. New York: Free Press.

Shea, N., & Frith, C. (2019). The global workspace needs metacognition. *Trends in Cognitive Sciences, 23*, 560–571.

Simon, N. W., Gilbert, R. J., Mayse, J. D., Bizon, J. L., & Setlow, B. (2009). Balancing risk and reward: A rat model of risky decision-making. *Neuropsychopharmacology, 34*(10), 2208–2217.

Siposova, B., Tomasello, M., & Carpenter, M. (2018). Communicative eye contact signals a commitment to cooperate for young children. *Cognition, 179*, 192–201.

Skinner, B. F. (1966). The phylogeny and ontogeny of behavior. *Science, 153*, 1205–1213.

Smaers, J. B., Gómez-Robles, A., Parks, A. N., & Sherwood, C. C. (2017). Exceptional evolutionary expansion of prefrontal cortex in great apes and humans. *Current Biology, 27*(5), 714–720.

Smith, J. D. (2009). The study of animal metacognition. *Trends in Cognitive Sciences, 13*(9), 389–396.

Smith, J. D., Schull, J., Strote, J., McGee, K., Egnor, R., & Erb, L. (1995). The uncertain response in the bottlenosed dolphin (*Tursiops truncatus*). *Journal of Experimental Psychology: General, 124*(4), 391–408.

Sterelny, K. (2001). *The evolution of agency and other essays.* Cambridge: Cambridge University Press.

Sterelny, K. (2004). *Thought in a hostile world.* Hoboken, NJ: Blackwell.

Stevens, J. R., Rosati, A. G., Ross, K. R., & Hauser, M. D. (2005). Will travel for food: Spatial discounting and reward magnitude in two New World monkeys. *Current Biology, 15*, 1855–1860.

Stiner, M. C. (2013). An unshakable Middle Paleolithic? Trends versus conservatism in the predatory niche and their social ramifications. *Current Anthropology, 54*(S8), S288–S304.

Suboski, M. D. (1992). Releaser-induced recognition learning by amphibians and reptiles. *Animal Learning and Behavior, 20*(1), 63–82.

Suddendorf, T., Crimston, J., & Redshaw, J. (2017). Preparatory responses to socially determined, mutually exclusive possibilities in chimpanzees and children. *Biology Letters, 13*, 20170170.

Szabo, B., Noble, D. W., Byrne, R. W., Tait, D. S., & Whiting, M. J. (2018). Subproblem learning and reversal of a multidimensional visual cue in a lizard: Evidence for behavioural flexibility? *Animal Behaviour, 144*, 17–26.

Szabo, B., Noble, D. W., Byrne, R. W., Tait, D. S., & Whiting, M. J. (2019). Precocial juvenile lizards show adult level learning and behavioural flexibility. *Animal Behaviour, 154*, 75–84.

Szabo, B., Noble, D. W., & Whiting, M. J. (2019). Context-specific response inhibition and differential impact of a learning bias in a lizard. *Animal Cognition, 22*, 317–329.

Szabo, B., Noble, D. W., & Whiting, M. J. (in press). Non-avian reptile learning 40 years on: Advances, promises and potential. *Biological Reviews.*

Szabo, B., & Whiting, M. J. (2020). Do lizards have enhanced inhibition? A test in two species differing in ecology and sociobiology. *Behavioural Processes, 172*, 104043.

Templer, V. L., Lee, K. A., & Preston, A. J. (2017). Rats know when they remember: Transfer of metacognitive responding across odor-based delayed match-to-sample tests. *Animal Cognition, 20*(5), 891–906.

Tennie, C., Völter, C. J., Vonau, V., Hanus, D., Call, J., & Tomasello, M. (2019). Chimpanzees use observed temporal directionality to learn novel causal relations. *Primates, 60*(6), 517–524.

Thomas, R. K. (1980). Evolution of intelligence: An approach to its assessment. *Brain, Behavior and Evolution, 17*(6), 454–472.

Todd, P. M., & Gigerenzer, G. E. (2012). *Ecological rationality: Intelligence in the world.* Oxford: Oxford University Press.

Tolman, E. C. (1948). Cognitive maps in rats and men. *Psychological Review, 55*(4), 189–208.

Tomasello, M. (2006). Why don't apes point? In N. Enfield & S. Levinson (Eds.), *Roots of human sociality: Culture, cognition and interaction* (pp. 506–524). Oxford: Berg.

Tomasello, M. (2008). *Origins of human communication.* Cambridge, MA: MIT Press.

Tomasello, M. (2014). *A natural history of human thinking.* Cambridge, MA: Harvard University Press.

Tomasello, M. (2016). *A natural history of human morality.* Cambridge, MA: Harvard University Press.

Tomasello, M. (2018). How children come to understand false beliefs: A shared intentionality account. *Proceedings of the National Academy of Sciences, 115*(34), 8491–8498.

Tomasello, M. (2020). The moral psychology of obligation. *Behavioral and Brain Sciences, 43*(e56), 1–58.

Tomasello, M., & Call, J. (1997). *Primate cognition.* New York: Oxford University Press.

Tomasello, M., & Carpenter, M. (2005). The emergence of social cognition in three young chimpanzees. *Monographs of the Society for Research in Child Development, 70*(1), vii–132.

Tomasello, M., Melis, A. P., Tennie, C., Wyman, E., & Herrmann, E. (2012). Two key steps in the evolution of human cooperation: The interdependence hypothesis. *Current Anthropology, 53*(6), 673–692.

Tomonaga, M., Myowa-Yamakoshi, M., Mizuno, Y., Okamoto, S., Yamaguchi, M., Kosugi, D., Bard, K., Tanaka, M., and Matsuzawa, T. (2004). Development of social cognition in infant chimpanzees (*Pan troglodytes*): Face recognition, smiling, gaze and the lack of triadic interactions. *Japanese Psychological Research, 46*, 227–235.

Tooby, J., & Cosmides, L. (1992). The psychological foundations of culture. In J. H. Barkow, L. Cosmides, & J. Tooby (Eds.), *The adapted mind: Evolutionary psychology and the generation of culture* (pp. 19–136). Oxford: Oxford University Press.

Tooby, J., & Cosmides, L. (2005). Conceptual foundations of evolutionary psychology. In D. Buss (Ed.), *The handbook of evolutionary psychology* (pp. 5–67). Hoboken, NJ: Wiley.

Turchin, P. (2016). *Ultrasociety: How 10,000 years of war made humans the greatest cooperators on earth.* Chaplin, CT: Beresta Books.

Vaish, A., Carpenter, M., & Tomasello, M. (2016). The early emergence of guilt-motivated prosocial behavior. *Child Development, 87,* 1772–1782.

Veissière, S., Constant, A., Ramstead, M., Friston, K., & Kirmayer, L. (2019). Thinking through other minds: A variational approach to cognition and culture. *Behavioral and Brain Sciences, 43*(e90), 1–75.

Völter, C. J., & Call, J. (2014). Younger apes and human children plan their moves in a maze task. *Cognition, 130*(2), 186–203.

Völter, C. J., Rossano, F., & Call, J. (2015). From exploitation to cooperation: Social tool use in orang-utan mother-offspring dyads. *Animal Behaviour, 100,* 126–134.

Völter, C. J., Sentís, I., & Call, J. (2016). Great apes and children infer causal relations from patterns of variation and covariation. *Cognition, 155,* 30–43.

von Uexküll, J. (1934/2010). *A foray into the worlds of animals and humans: With a theory of meaning* (J. D. O'Neil, Trans.). Minneapolis: University of Minnesota Press.

von Uexküll, J. (1909). *Umwelt und Innenwelt der Tiere.* Berlin: Springer.

Walsh, D. M. (2015). *Organisms, agency, and evolution.* Cambridge: Cambridge University Press.

Warneken, F., Lohse, K., Melis, A. P., & Tomasello, M. (2011). Young children share the spoils after collaboration. *Psychological Science, 22*(2), 267–273.

Wiener, N. (1948). *Cybernetics: Or control and communication in the animal and the machine.* Cambridge, MA: MIT Press.

Wilkinson, A., & Huber, L. (2012). Cold-blooded cognition: Reptilian cognitive abilities. In J. Vonk & T. K. Shackelford (Eds.), *The Oxford handbook of comparative evolutionary psychology* (pp. 129–143). Oxford: Oxford University Press.

Wilson, D. S., & Wilson, E. O. (2007). Rethinking the theoretical foundation of sociobiology. *Quarterly Review of Biology, 82*(4), 327–348.

Wilson, E. O. (1975). *Sociobiology: The new synthesis.* Cambridge, MA: Harvard University Press.

Wilson, E. O. (2012). *The social conquest of earth*. New York: W. W. Norton.

Winterhalder, B., & Smith, E. A. (2000). Analyzing adaptive strategies: Human behavioral ecology at twenty-five. *Evolutionary Anthropology: Issues, News, and Reviews, 9*(2), 51–72.

Wittgenstein, L. (1921). *Tractatus logico-philosophicus*. Abingdon: Routledge.

Wolf, W., & Tomasello, M. (2020a). Human children, but not great apes, become socially closer by sharing an experience in common ground. *Journal of Experimental Child Psychology, 199*, 104930.

Wolf, W., & Tomasello, M. (2020b). Watching a video together creates social closeness between children and adults. *Journal of Experimental Child Psychology, 189*, 104712.

Yin, H. H., & Knowlton, B. J. (2006). The role of the basal ganglia in habit formation. *Nature Reviews Neuroscience, 7*(6), 464–476.

Yu, A. J., and Dayan, P. (2005). Uncertainty, neuromodulation, and attention. *Neuron, 46*, 681–692.

Index

Page numbers in italics indicate figures.